Foundations for Interdisciplinarity in the Life Sciences: Concise Monographs

The Foundations for Interdisciplinarity in the Life Sciences book series introduces and extends scholarship addressing a growing recognition that biological phenomena which suggest agency, directionality, and goal-directedness demand new or updated conceptual frameworks that can translate into rigorous theoretical models and discriminating empirical tests. The book series provides readers with conceptual, theoretical, and experimental resources for prosecuting new kinds of inquiry across disciplinary boundary lines.

Álvaro Moreno • Juli Peretó

An Evolutionary Story of Agency

How Life Evolved to Act on its Own

Springer

Álvaro Moreno
University of Basque Country, Donostia
International Physics Center
San Sebastian, Spain

Juli Peretó
Institute for Integrative Systems Biology
I2SysBio (University of Valencia-CSIC)
and Department of Biochemistry and
Molecular Biology
University of Valencia
Paterna, Spain

ISSN 3005-1231 ISSN 3005-124X (electronic)
Foundations for Interdisciplinarity in the Life Sciences: Concise Monographs
ISBN 978-3-032-05643-6 ISBN 978-3-032-05644-3 (eBook)
https://doi.org/10.1007/978-3-032-05644-3

This work was supported by John Templeton Foundation

This Springer imprint is published by the registered company Springer Nature Switzerland AG.
The registered company address is: Gewerbestrasse 11, 6330 Cham, Switzerland

Acknowledgments

AM extends heartfelt gratitude to Kepa Ruiz-Mirazo and Leonardo Bich from the IAS Research Group, as well as Ben Shirt-Ediss from DIPC, San Sebastián, Spain, for their collaboration, critical feedback, and insightful suggestions on various parts of the manuscript. Special thanks also go to Matteo Mossio (IHPST, Paris) for valuable discussions that contributed to this research. Of particular significance was the collaboration with Argyris Arnellos from the University of the Aegean, Syros (Greece), whose contributions were instrumental in shaping this work.

AM is especially grateful to Kepa Ruiz-Mirazo for his unwavering support and encouragement throughout this research process. Without his guidance, this book would not have been possible.

Acknowledgment is also due to the John Templeton Foundation for funding support through research project 62220, "Agency, Directionality & Function—Science of Purpose." This support provided not only financial resources but also an intellectually stimulating environment for developing this research. In particular, AM would like to express deep appreciation to Alan Love for his continuous and generous support during this period.

A special note of thanks is extended to the Donostia International Physics Center (DIPC), as well as to its president and founder, Pedro M. Etxenike, and its director, Ricardo Díez Muiño. The administrative and operational facilities provided by the DIPC were essential to the realization of this book.

JP wishes to thank AM for the opportunity to engage in this fascinating discussion and for the many hours of enriching dialogue between a philosopher and a biochemist. He also acknowledges the welcoming and stimulating scientific environment of the Integrative Systems Biology Institute (I^2SysBio, Universitat de València-CSIC), where he is enjoying the final years of his academic career.

The authors apologize in advance and acknowledge that there is a vast bibliography on the subject that cannot be reflected in full, given the space limitations of this monograph and the fact that it is not an exhaustive review but rather a reflection on current scientific knowledge on such an open-ended issue.

Both authors would also like to express their gratitude to Iñaki Ruiz-Trillo and Elena Casacuberta from the Institut de Biologia Evolutiva (CSIC-UPF, Barcelona) for their inspiring discussions on the origin of metazoans, and to thank Leonardo Bich, Eva Jablonka, Kepa Ruiz-Mirazo and two anonymous reviewers for their careful and critical reading. Any errors that may remain in the text are the sole responsibility of the authors.

Special thanks go to Jaume Güemes, director of the Jardí Botànic de la Universitat de València, for generously providing research facilities during AM's stays in Valencia in 2024 and 2025—at both the beginning and the conclusion of this journey.

Competing Interests The authors have no competing interests to declare that are relevant to the content of this manuscript.

Contents

Chapter 1
Introduction

Abstract Traditional philosophical perspectives have often linked agency to human rationality and intentionality, but biological evidence suggests that agency predates human cognition and manifests in simpler life forms. This chapter explores agency as a biological phenomenon, examining its evolutionary origins and its role in shaping adaptive behavior. By investigating major evolutionary transitions this work frames agency as an integral driver of life's increasing complexity. The study challenges conventional views by arguing that agency is not merely a product of evolution but an active force influencing evolutionary trajectories. Ultimately, this approach underscores the dynamic interplay between biological organization, environmental interactions, and the emergence of complex behaviors, offering a framework for understanding the evolution of agency beyond human cognition.

Keywords Agency · Major evolutionary transitions · Biological complexity · Goal-directed behavior · Agency complexification

The concept of agency has attracted increasing attention across disciplines such as biology, robotics, and psychology. However, its complexity and the diversity of interpretations it has generated present substantial challenges. The terms agent and agency are used to describe a wide range of phenomena, depending on the field. For roboticists, agency often refers to behaviors not explicitly pre-programmed by humans. In computer science, agents are typically software entities capable of adapting and performing tasks in dynamic environments. Biologists understand agency as adaptive behavior exhibited by organisms or collectives such as colonies. In psychology, agency is frequently associated with intentional action. This conceptual diversity reflects not only disciplinary differences but also deeper philosophical disagreements, complicating efforts to clarify the nature of agency itself.

According to Webster's Dictionary, agency is the capacity to act or exert power, while the Stanford Encyclopedia of Philosophy defines it as the capacity to exercise

Á. Moreno, J. Peretó, *An Evolutionary Story of Agency*, Foundations for Interdisciplinarity in the Life Sciences: Concise Monographs,
https://doi.org/10.1007/978-3-032-05644-3_1

direct control or guidance over one's own behavior. Both definitions emphasize that agency involves entities—agents—with causal powers to carry out actions. Implicit in this idea is the notion of goal-directedness: to act with control is, by definition, to act with purpose. In broad terms, then, agency can be understood as the capacity of a system to produce actions that originate from within and are directed toward achieving specific goals.

Historically, the study of agency has been dominated by philosophy, with an emphasis on complex human agency—deliberate, future-oriented behavior shaped by beliefs, desires, intentions, and rational planning. Yet many forms of goal-directed behavior in the biological world occur without conscious deliberation. Animals exhibit a vast repertoire of adaptive behaviors—fleeing, nesting, communicating—that help maintain internal order, group cohesion, or reproductive success in the face of changing environments. These behaviors are marked by flexibility and responsiveness, which are core features of agency.

Moreover, agency is not limited to animals. Sessile organisms like plants and fungi also engage in adaptive interactions with their environments. Plants secrete chemicals to deter herbivores or to signal other organisms; fungi release enzymes to digest and absorb nutrients or to interact symbiotically with plants. Across the biological world, organisms act to modify their surroundings: they move toward favorable conditions, acquire resources, avoid threats, reproduce, and communicate chemically or behaviorally. These goal-oriented, adaptive interactions provide a broad, functional definition of agency.

In biology, then, agency can be understood as a phenomenon in which an individuated organism becomes the source of interactions that alter its environment to produce functional outcomes—specifically, effects that promote its own maintenance or reproductive continuity.

Approaching agency from an evolutionary perspective—beginning with its simplest manifestations—allows us to better understand the biological foundations of more complex forms, including human agency. This perspective reframes agency not as an exclusively human trait, but as a biologically grounded phenomenon with deep evolutionary roots. By tracing the emergence and development of agency over evolutionary time, we can identify the mechanisms that have shaped its increasing complexity and its role in adaptive behavior across species.

Our work aims to develop a comprehensive theory of the origins and evolution of agency within a biological framework. Such a theory explores how agency emerged, evolved, and became more complex through critical evolutionary milestones. But the goal is not simply to recount a historical progression. Rather, the book seeks to uncover the underlying causes and broader consequences of agency. Starting with the emergence of the first cells, it investigates how early life forms developed basic goal-directed behaviors to influence their surroundings. From this foundation, it examines how certain lineages evolved increasingly complex actions, offering a causal reinterpretation of the evolutionary trajectory of agency.

Although many researchers regard agency as a defining feature of life, its development has been uneven. Plants and fungi, for instance, display forms of agency that differ greatly from the highly integrated behaviors seen in mammals. This disparity raises important questions about the role of agency in the history of life and the factors that shaped its selective elaboration. It may be that agency became more complex through contingent innovations in certain multicellular organisms, opening new evolutionary pathways. Once established, agency may have driven further complexity, leading to new forms of integration and adaptation.

This book proposes a novel thesis: to understand the history of life, we must construct a unified theory of agency—one that synthesizes empirical evidence to explain its evolutionary dynamics. Such a theory must go beyond cataloging mechanisms; it must also identify general principles, address fundamental questions, and account for both the causal and teleological dimensions of agency. That is, it must explain how organisms act in ways that preserve their own existence and influence their evolutionary trajectories.

The evolution of agency is not a linear process. It involves a dynamic interplay of historical contingencies, simultaneous interactions, and functional outcomes nested within complex causal loops. Understanding this interplay is essential for grasping the biological nature of agency—and for making sense of life's history itself.

1.1 Major Evolutionary Transitions

Throughout Earth's history, life has followed diverse evolutionary trajectories. While many unicellular organisms have remained unchanged in form, others evolved multicellularity, opening new evolutionary pathways. Change is inherent to life—organisms cannot persist without it. But what drives these changes?

Increasingly, it is recognized that evolution is not merely the result of "external" selective pressures acting upon passive organisms. Rather, organisms are active participants in their own evolution. Their organizational architecture enables them to explore viable possibilities under varying ecological conditions. Occasionally, profound organizational shifts occur. If these changes are viable, they can open new evolutionary paths, sometimes enabling the occupation of previously inaccessible ecological niches. In this sense, organisms—and their activities—actively shape the environments they inhabit, as well as the selective pressures that influence their evolutionary trajectories.

This drive to change lacks any pre-defined direction and has produced a multitude of evolutionary outcomes. Evolution has explored all viable avenues in response to local conditions. Some of these changes have led to novelty—occasionally toward greater complexity, often toward diversification within a given structural framework, and at times even toward simplification.

Yet, a small fraction of these paths has resulted in a causally sustained trajectory of increasing complexity. The existence of such a trajectory does not imply that evolution is inherently directional or that organisms are selected for complexity. Simpler life forms remain the vast majority, and complexification has taken many forms. Each step toward complexity has spawned countless branches, with only a few continuing along that path. Nevertheless, viewed globally, the empirical evidence shows that one evolutionary path has indeed proceeded through successive levels of complexity.

At each major transition, some of the diverse forms that emerged laid the foundation for further leaps in complexity. Evolution, then, is a vast exploration of possibilities—adapting to each new level of organization, occasionally breaching thresholds of complexity, and frequently encountering ceilings that constrain further elaboration. Examples include the transitions from prokaryotic to eukaryotic cells and the emergence of metazoans from unicellular protists. These major innovations, opening new realms of exploration, mark critical shifts in evolutionary potential.

"Major Evolutionary Transitions" refer to significant qualitative shifts in evolution. John Maynard Smith and Eörs Szathmáry explored this concept, linking it to the emergence of new modes of information transmission and the transformation of simpler entities into more complex individualities (Fig 1.1). Their framework has inspired growing interest while also sparking debates over how major transitions should be defined. Some, like Robin et al. (2021), advocate for a broader, ecosystem-level perspective, while others, such as Okasha (2024), question their ontological and explanatory status.

Today, a major evolutionary transition is generally understood as the process by which a qualitatively new phenomenological domain emerges in the history of life, rather than merely the formation of new units of selection from pre-existing entities. Although this remains a debated issue, we argue that the key factor in determining whether a transition qualifies as "major" is the magnitude of its evolutionary consequences, particularly if these are linked to profound changes in the individual entities involved in the process. This perspective, however, makes it challenging to categorize all purported transitions under a single universal criterion.

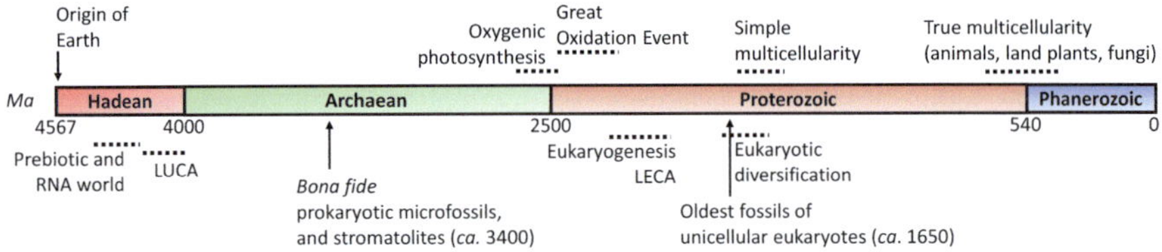

Fig. 1.1 Major transitions in evolution. According to Maynard Smith and Szathmáry (1997), biological evolution has undergone transformative shifts through new modes of information transmission, enabling simpler entities to evolve into more complex individualities. Examples of such transitions include the emergence of the genetic code, meiotic sex, and language. In this work, we focus on the expansion of agential capabilities across key evolutionary milestones: the origin of life (Chap. 2), prokaryotic diversification (Chap. 3), eukaryogenesis and protist evolution (Chap. 4), the evolution of simple and complex (true) multicellularity (Chaps. 4 and 5), and the emergence of metazoans and nervous systems (Chap. 6). The figure presents approximate dates for some of these major events, based on data from Cantine and Fournier (2018), Javaux (2019), and Miao et al. (2024). LUCA (Las Universal Common Ancestor), LECA (Last Eukaryotic Common Ancestor)

The concept of major evolutionary transitions suggests that certain evolutionary pathways result in significant increases in complexity, though complexity itself is hard to define precisely. While life's history is full of contingencies, evolution's vast timescale and spatial reach have allowed the exploration of many possibilities, leading to repeated patterns. For example, different degrees of multicellularity have evolved independently at least 25 times. New research indicates evolutionary trajectories are not entirely random and are constrained by finite physico-genetic and ecological motifs. These constraints shape basic morphological designs and ecosystem architectures, limiting evolutionary possibilities. Thus, while contingencies exist, the emergence of certain patterns may reflect deeper organizing principles that govern life's evolutionary history.

The remarkable increase in complexity along certain evolutionary paths raises the question of whether biological organization itself has actively contributed to this process. While many argue that complexity arises solely from external forces such as selective pressures and environmental changes, others contend that organisms play an active role in evolution. An organism's phenotype results from hierarchical interactions within its developmental processes, where genetic changes are constrained to specific regions within the morphospace. These processes follow an intrinsic logic, often producing changes that are not directly adaptive but stem from complex internal adjustments. As a result, organisms can generate novel traits that later become subject to natural selection. This perspective suggests that evolution is less about random exploration and more about progression along predetermined paths. Thus, the focus shifts from understanding how evolution creates complex organisms to exploring how the active role of organisms influences evolutionary trajectories. This framework highlights the interplay between internal biological mechanisms and external pressures, offering a nuanced view of evolutionary complexity.

1.2 Understanding Evolution Through Agency

Organisms maintain themselves by actively modifying their environment. They move toward more favorable conditions to find food, avoid predators, or locate suitable places to reproduce. They secrete chemicals to communicate, deter enemies, or successfully invade hosts. In doing so, organisms actively construct their own niches and, to a significant extent, regulate how they adapt over time—within the span of their own lifetimes. This perspective shifts the traditional question from *how evolution produces complex organisms* to *how the active role of organisms themselves might help explain the different evolutionary trajectories observed throughout life's history*.

One of the most remarkable features of living beings is their ability to use their internal organization to act upon the environment, including other organisms. Living beings perform actions—purposeful interactions with their surroundings. They detect changes in their environment and possess the capacity to modify them. When we speak of an organism's actions, we mean that its organization imposes functional constraints on the environment. These constraints effectively extend the organism's internal structure outward, generating processes that contribute to its own maintenance and reproductive continuity. For instance, the metabolic organization of a bacterium regulates flagellar movement, propelling it toward higher concentrations of sugar, which it then absorbs to sustain its metabolism.

Such actions are crucial for survival, as their outcomes allow organisms to obtain resources, evade threats, and ultimately reproduce. Action entails anticipation: the organism enacts a change in the environment (the effect) in response to a need or stimulus (the cause). In turn, this external change triggers an internal adjustment—for example, the restoration of homeostasis. In animals, neurophysiological stability depends on the constant performance of such actions, the majority of which must succeed to preserve life.

Organisms adjust their actions based on environmental conditions in order to maintain their identity. This is why these actions are described as "goal-directed" or "purposeful." Importantly, such goal-directedness does not require conscious thought. It simply means that the organism is structured in such a way that it reliably carries out certain actions critical to its continued existence. If it fails to do so, its organization—and therefore its life—will cease.

However, this definition is intentionally broad, as the range of adaptive interactions organisms engage in is extremely diverse. Not all such interactions may qualify strictly as "agentive." In some cases, organisms respond to environmental stimuli by altering only their internal organization. In others, they may excrete substances that influence their surroundings in ways that are ecologically significant only over long timescales and across vast populations. These examples do not align neatly with conventional definitions of agency, which typically imply actions directed toward goals immediately relevant to the agent's own survival or functioning.

Moreover, agency in biology is not always limited to individual organisms. It often emerges from collective systems—such as colonies, symbiotic partnerships, or other cooperative arrangements—creating distributed or collective forms of agency. The complex interplay between individual organisms and these larger-scale systems introduces new layers of causality and function, making it necessary to refine our understanding of how organisms adaptively and causally modify their environments.

To make sense of this often paradoxical diversity, we will explore the role of agency in the history of life. We propose that agency is central to understanding the wide variety of evolutionary paths and, in particular, to explaining certain major evolutionary transitions. This view builds on the idea that every living organism is a complex, self-constituting, and self-sustaining system. The survival of an organism—both as an individual and as part of a lineage—depends on its capacity to act effectively. If successful action is a prerequisite for survival and reproduction, then it becomes a key factor in shaping evolution. The type of agency exhibited by different lineages will vary significantly, and these different forms of agency will have lasting evolutionary consequences.

As a result, the behavior of organisms in their environments should not be viewed merely as outcomes to be explained by evolution (explananda), but also as explanatory elements (explanans)—forces that drive evolutionary change. What we need, then, is a concept of agency that maintains a basic conceptual continuity across the many branches of the tree of life. For the purposes of this inquiry, we do not begin with a highly elaborate or definitive concept. Rather, our aim is to let the research itself guide us toward a more refined understanding. Accordingly, we inform the reader that a more detailed and precise characterization of agency will be developed and presented later in the book.

1.3 Understanding Agency Through Its Evolution

We have briefly presented a view of life in our planet, asking ourselves what are the causes of one of the most intriguing facts—and at the same time, most impressive—of the phenomenon of life: its ability to diversify and, above all, to become more and more complex. We have also argued why agency plays an active role in the evolution of life, thus linking evolution with the subject of this book. Now, we will discuss why the question of how the diverse forms of agency of organisms play a causal role in the complexification of life and how agency could be related with the notion of major evolutionary transitions.

Life is equally successful in the simple agentive forms of bacteria, plants or fungi than in the extraordinary display of agency in animals. Yet, not all forms of agency play the same explanatory role in the evolution and complexification of life. What organisms must do to maintain themselves alive varies greatly, depending on their ways of life, size and other factors. For example, some organisms have evolved in a way that they need to deploy movement, while others have been adapted successfully to forms involving a sessile style of life and their agency relies mostly on chemical secretions.

How can we understand the significance of how organisms engage functionally with their environment and other living systems—a capability that was crucial even in the earliest forms of life? This leads to a deeper question: why has agency become more complex in only certain evolutionary branches? The answer may lie in how we conceptualize the relationship between life and agency. It might be more accurate to propose that, by chance, the appearance of a particular form of multicellular life provided the conditions for a new type of agency to emerge. Once this opportunity arose and was exploited, this new agential dimension could flourish, paving the way for more complex and integrated forms of life.

For instance, some multicellular organisms remained sessile, while others developed a way of life centered on macroscopic motile agency, which introduced further innovations such as vision-guided flight. These divergent pathways illustrate the selective unlocking of opportunities for agential complexity. Addressing these questions is essential for understanding the history of life: why and how have only specific evolutionary trajectories allowed for the increasing complexity of agency?

As we will explore, the continuity of organisms is closely tied to their form of agency. Variations in the type and complexity of agency significantly influence their evolutionary trajectories. If there is a correlation between the complexity of agency and that of the organism, then the emergence of more complex forms of agency will play a causal role in sustaining more complex organisms. This is because maintaining organismic complexity relies on the success of their behaviors. Therefore, the complexification of agency serves as an explanans for the specific evolutionary paths that have driven the increasing complexity of organisms.

Our research aims to examine not only the efficient causes behind each step in life's history but also the success or failure of these steps based on their outcomes. Organisms do things *in order to achieve certain functional goals* (getting food, escaping a predator), the achievement of which determines their survival and, ultimately, the evolutionary continuity of their organizational identity. What they do "for" is therefore a relevant explanatory factor. The history of life arises from a complex interplay of factors: events are shaped by prior conditions, contemporaneous influences, and the ability of organisms to create nested causal loops through their functional roles.

Adopting a teleological perspective, we highlight the functional significance of key processes. Our investigation into agency in the history of life seeks to uncover how, during primordial biogenesis, some systems developed the capacity to purposefully modify their environments and perform specific functions. We then analyze why, over the course of evolution, certain forms of agency became progressively more complex.

Agency is intrinsically linked to organisms' physiological, ecological, and evolutionary roles as integrated biological systems. Understanding its evolution involves explaining how complex agents emerged and sustained themselves. Agency, inherently teleological, operates to maintain life, enabling survival, reproduction, and adaptation. This goal-driven causality extends beyond individuals to influence long-term evolutionary processes, driving population and ecosystem regeneration and fostering innovation. In turn, the emergence of complex agency—the phenomenon we seek to explain—also serves as a causal force in life's evolution.

The tree of life reflects the forms of agency exhibited by its last common ancestors. Distinct evolutionary paths have unlocked varying degrees of complexity, shaping diverse possibilities and constraints. Some developmental architectures facilitated the evolution of advanced capacities for outward action, while others reached limits and pursued alternative trajectories. Environmental conditions have influenced these outcomes, but life itself has actively transformed the biosphere and the planet.

1.4 What This Book Is About

This book is a conceptual exploration of agency, addressing questions at the intersection of theoretical biology, cognitive science, philosophy, artificial intelligence, and synthetic biology. Its interdisciplinary scope demands a methodological approach that integrates philosophical analysis with empirically grounded scientific theories. Rejecting speculative reasoning, the book aims to propose testable frameworks, interweaving philosophical reflection with scientific investigation to enhance understanding of agency's role in life's evolution and the potential creation of artificial agents.

In this context, it is timely to ask whether insights from biological agency can inform artificial systems, or vice versa. Biological organisms, uniquely capable of autonomy and agency, derive these traits from their material organization. While synthetic biology manipulates biological systems, it extends the study of biological agency rather than creating truly artificial forms. Developing autonomous agents from inert materials remains a major challenge, as true agency depends on a level of autonomy that such systems have yet to achieve. Robots, for instance, are not the product of processes like self-construction, self-repair, or self-maintenance. They lack a metabolic system whose failure would result in their demise and the absence of offspring. For these reasons, studying agency within biological systems offers deeper insights than focusing exclusively on artificial or robotic models.

Nonetheless, advancements in artificial intelligence offer new possibilities. Sophisticated computational networks trained on human intelligence are forming hybrid systems capable of collective agency. These developments, though outside the book's main focus, highlight potential forms of non-biological agency, raising significant questions for future exploration.

This book emphasizes the evolutionary trajectory of agency, particularly its origins, functions, and increasing complexity, tracing its development up to the emergence and evolution of metazoans with nervous systems. While the mind—a key aspect of complex agency—is acknowledged, this work concentrates on the early evolution of agency, culminating in a transformative stage marked by the advent of mental agency. In this work, thus, we focus on the expansion of agential capabilities across key evolutionary milestones: the origin of life (Chap. 2), prokaryotic diversification (Chap. 3), eukaryogenesis and protist evolution (Chap. 4), the evolution of simple and complex multicellularity (Chaps. 4 and 5), and the emergence of metazoans and nervous systems (Chap. 6), a pivotal transition revealing the appearance of mind as both a product and driver of agency's evolution (Fig. 1.1).

Instead of general terms like "basic" or "cognitive" agency, the book proposes a precise framework linking agency's evolution to major transitions in life's history. It underscores agency's active role in shaping evolutionary processes, creating a "virtuous explanatory circle" where agency and evolution mutually enhance understanding. By examining these dynamics, this research aspires to provide new insights into the nature of agency, its evolutionary significance, and its central role in life's unfolding narrative. Ultimately, we hope to advance scientific understanding by revealing how the evolution of agency and the dynamics of life are deeply interconnected.

Bibliography

Alberch P (1982) Developmental constraints in evolutionary processes. In: Bonner JT (ed) Evolution and development. Dahlem Workshop Reports, vol 22. Springer, Berlin, Heidelberg, pp 313–332. https://doi.org/10.1007/978-3-642-45532-2_15

Ball P (2023) Organisms as agents of evolution. John Templeton Foundation, West Conshohocken. https://www.templeton.org/wp-content/uploads/2023/04/Biological-Agency_1_FINAL.pdf

Ball P (2024) How life works. A user's guide to the new biology. Picador, London

Barandiaran X, di Paolo E, Rohde M (2009) Defining agency: individuality, normativity, asymmetry, and spatio-temporality in action. Adap Behav 17:367–386. https://doi.org/10.1177/1059712309343819

Burge T (2009) Primitive agency and natural norms. Philos Phenom Res 79:251–278. http://www.jstor.org/stable/40380466

Cantine MD, Fournier GP (2018) Environmental adaptation from the origin of life to the last universal common ancestor. Orig Life Evol Biosph 48:35–54. https://doi.org/10.1007/s11084-017-9542-5

Frankfurt HG (1978) The problem of action. Am Philos Q 15:157–162. https://www.jstor.org/stable/20009708

Godfrey-Smith P (2017) The subject as cause and effect of evolution. Interface Focus 7:20170022. https://doi.org/10.1098/rsfs.2017.0022

Javaux EJ (2019) Challenges in evidencing the earliest traces of life. Nature 572:451–460. https://doi.org/10.1038/s41586-019-1436-4

Juarrero A (1999) Dynamics in action: intentional behavior as a complex system. MIT Press, Cambridge, MA. https://doi.org/10.7551/mitpress/2528.001.0001

Lala KN, Uller T, Feiner N et al (2024) Evolution evolving: the developmental origins of adaptation and biodiversity. Princeton University Press, Princeton, NJ

Love A, Dresow M (2022) Organizing interdisciplinary research on purpose. BioScience 72:321–323. https://doi.org/10.1093/biosci/biab128

Maynard-Smith J, Szathmáry E (1997) The major transitions in evolution. Oxford University Press, Oxford

Miao L, Yin Z, Knoll AH et al (2024) 1.63-billion-year-old multicellular eukaryotes from the Chuanlinggou Formation in North China. Sci Adv 10:eadk3208. https://doi.org/10.1126/sciadv.adk3208

Moreno A, Etxeberria A (2005) Agency in natural and artificial systems. Artif Life 11:161–176. https://doi.org/10.1162/1064546053278919

Moreno A, Mossio M (2015) Biological autonomy: a philosophical and theoretical enquiry. Springer, Dordrecht. https://doi.org/10.1007/978-94-017-9837-2

Okasha S (2024) The concept of agent in biology: motivations and meanings. Biol Theory 19:6–10. https://doi.org/10.1007/s13752-023-00439-z

Robin AN, Denton KK, Horna Lowell ES et al (2021) Major evolutionary transitions and the roles of facilitation and information in ecosystem transformations. Front Ecol Evol 9:711556. https://doi.org/10.3389/fevo.2021.711556

Solé R, Kempes CP, Corominas-Murtra B et al (2024) Fundamental constraints to the logic of living systems. Interface Focus 14:20240010. https://doi.org/10.1098/rsfs.2024.0010
Skewes J, Hooker C (2009) Bio-agency and the problem of action. Biol Philos 24:283–300. https://doi.org/10.1007/s10539-008-9135-9
Virenque L, Mossio M (2024) What is agency? A view from autonomy theory. Biol Theor 19:11–15. https://doi.org/10.1007/s13752-023-00441-5
Walsh DM (2015) Organisms, agency and evolution. Cambridge University Press, Cambridge, UK

Chapter 2
Minimal Agents and the Emergence of Life

Abstract This chapter explores the origins of agency in prebiotic systems, highlighting the progression from simple protocells to complex adaptive protocells. These early systems exhibited self-maintenance and adaptive interactions with their environments, leading to increased complexity and the development of functional action-detection loops. By examining the evolution from basic self-producing and reproducing systems to more advanced adaptive protocells, we try to elucidate how early life forms developed mechanisms to control and respond to environmental changes, paving the way for the emergence of primitive colonial systems.

Keywords Prebiotic systems chemistry · Prebiotic evolution · Protocells · Minimal agency

2.1 Emergence of Life: Hypotheses, Experiments, Simulations, Concepts

The natural emergence of life on Earth can be understood as a logical outcome derived from evolutionary theory. Both Lamarck and Darwin envisioned a chemical process giving rise to complex systems capable of further evolution through spontaneous generation. In modern terms, we embrace a materialistic and evolutionary explanation for the transition from geochemistry to biochemistry under early Earth conditions. As soon as a stable hydrosphere formed, chemical ingredients from terrestrial sources (such as atmospheric and volcanic chemistry) or extraterrestrial materials (including cosmic dust, asteroids, and comets) interacted to generate the first prebiotic self-organized systems operating far from thermodynamic equilibrium.

Today, various empirical and computational methods are employed to study the early stages of chemical evolution. Prebiotic chemistry, grounded in the Oparin-Haldane hypothesis of a "primordial soup" and advanced by Miller's iconic

Á. Moreno, J. Peretó, *An Evolutionary Story of Agency*, Foundations for Interdisciplinarity in the Life Sciences: Concise Monographs,
https://doi.org/10.1007/978-3-032-05644-3_2

experiments in the 1950s, sought to robustly synthesize biological building blocks under presumed primitive Earth conditions. For instance, Miller demonstrated the formation of amino acids using electrical discharges and gas mixtures simulating an early atmosphere, while Oró, in the 1960s, achieved the abiotic synthesis of the nitrogenous bases found in nucleic acids. Over decades, laboratories have replicated nearly the full range of biomolecules under abiotic conditions. Simultaneously, the analysis of organic material in meteorites has reinforced the idea that cosmochemistry enriched the prebiotic inventory. However, some essential biomolecules, particularly nucleotides, have proven challenging to synthesize from their basic components (nucleobases, sugars, and phosphates). The chemical processes involved in the abiotic synthesis of these components were often incompatible, resulting in intractable mixtures, or tars.

In recent years, systems chemistry has enabled the simultaneous synthesis of multiple biomolecular components in heterogeneous mixtures. In these systems, reactants may act as catalysts, and deterministic chemical forces seem to play a significant role. This aligns with the classical debate on chance versus necessity in evolution, suggesting that protometabolic chemistry—responsible for life's origin—is a specialized subset of broader abiotic chemistry. Thus, a restricted repertoire of molecules formed the foundation for the complex structures characteristic of life.

Recent progress in understanding non-enzymatic chemical transformations has highlighted the potential for organized reaction chains and catalytic cycles, offering a simplified model of metabolic networks. While key phenomena such as autocatalytic cycles remain elusive beyond a few abiotic examples, optimism about studying and reproducing protometabolic networks has grown.

The origin of life also involves physics, as biophysical principles are essential to understanding life's emergence and function across scales. These principles trace back to early chemical evolution, focusing on the emergence of self-organized systems that remain far from equilibrium by coupling matter and energy flows. Self-assembly processes, driven by entropic forces, were likely crucial from the start. Examples include the spontaneous formation of vesicles by lipid molecules in water and the folding of macromolecules. Additionally, nanoscale effects and size-dependent phenomena must be accounted for understanding life's origins.

The possible environments in which life could have emerged on the primitive planet have been the subject of controversy. Some prebiotic models propose that abiotic processes took place in surface waters with the participation of ultraviolet radiation and the chemical and physical effects associated with meteorite impacts. Other proposals emphasize the role of chemical and physical gradients associated with alkaline hydrothermal vents. If we adopt an eclectic and non-dogmatic position, we should not rule out the influence and contribution of any geochemical environment relevant to prebiotic chemistry.

Life as we know it relies on the interplay of two types of macromolecules: nucleic acids and proteins. These molecules are interdependent: nucleic acids (genomes) store the information required for protein synthesis, while proteins enable the replication and expression of nucleic acid information. This interdependence posed a significant challenge to understanding life's origins. A plausible solution was the hypothesis of a single type of macromolecule capable of both informational and catalytic functions, leading to the proposal of RNA as this dual-role molecule. The discovery of RNA catalysis (ribozymes) further strengthened the RNA world hypothesis, which posits that an evolutionary stage existed where RNA served both as genetic material and as a catalyst for its replication. Although systems chemistry suggests more complex scenarios involving the simultaneous emergence of RNA and DNA, the RNA world remains the prevailing model for explaining the transition to genetically instructed protein synthesis and the eventual replacement of RNA by DNA as genetic material.

From an empirical perspective, systems chemistry explores not only the properties of complex chemical mixtures and their transformations but also the behavior of diverse molecular components. This includes self-reproducing lipid vesicles containing RNA polymers. The assembly of simple protocells—vesicles equipped with self-replicating RNA, ribozymes for RNA monomer synthesis, and membrane components—appears to be a realistic and achievable milestone in the foreseeable future.

2.2 A Hypothetical Scenario for Biogenesis

Let us consider a scenario in which the conditions were conducive to the formation of countless heterogeneous assemblies of self-organizing microscopic vesicles. Some of these vesicles were coupled to chemical autocatalytic sets, creating mutual influences that enriched their dynamics. Chemical reactions intrinsic to these systems affected both the internal aqueous core and the physical properties of their physical boundaries—such as membrane permeability and fluidity. These compartmentalized systems likely engaged in aggregative and competitive interactions, undergoing transformations driven by environmental factors such as changes in osmolarity, pH, ion concentrations, temperature, and local fluid flow. These fluctuations in composition could have scaffolded primitive division cycles. Simultaneously, fusion events facilitated the redistribution and combination of vesicles from different sets, promoting diversity.

We refer to these self-assembling vesicles associated with chemical reactions as protocells. Over time, some protocells may have developed a cyclic, far-from-equilibrium form of self-maintenance—a spatially enclosed cyclic organization of material components and transformation processes that produced the components forming their physical boundaries. These boundaries, in turn, regulated the transport of energy and matter, sustaining the system. Such systems can be described as *autopoietic or self-producing protocells*—namely, protocells that harbor a network of reactions that contribute to maintain their physical compartment. Some of these early protocells might have evolved molecular autocatalytic mechanisms, enabling them to channel growth and reproduce, thereby generating highly similar entities. This marked the beginning of intergenerational continuity, fostering evolving populations subject to primitive selection or competition based on stability and robustness. These populations, in turn, interacted with and altered their environment, facilitating the emergence of primitive food webs—an essential development for the long-term maintenance of protocell populations.

Thus, primordial biogenesis introduced two distinct but interrelated processes: individual protocell (protometabolic) processes and collective (historical) processes that supported 'ecopoiesis' (i.e. the formation of the first ecosystems). These processes operated at different spatial and temporal scales but influenced one another through inter-level feedback loops. The emergence of more complex protocells with efficient reproductive capacities reinforced selective dynamics within evolving lineages, fostering the diversification of collective food webs. These collective, long-term processes, in turn, likely facilitated the organization and complexification of individual protocells, favoring the appearance of protocells with genetically instructed systems (e.g., RNA-driven metabolisms). This genetic instruction enabled reliable inheritance and phenotypic diversity, further strengthening selective dynamics within lineages and contributing to the ecological complexity of collective networks. In sum, this interplay laid the groundwork for subsequent evolutionary milestones on the path toward life.

The emergence and evolution of protocells were rooted in a primitive network of chemical reactions, initially driven by spontaneous reactions and simple catalytic processes, such as mineral catalysis or organometallic compounds. This protometabolic network supplied the basic components of protocells, such as membrane-forming lipids. The presence of metastable molecules—those with high energetic potential but kinetic stability—was crucial for the development of the first autocatalytic cycles, which sustained protocell growth. This protometabolic framework also served as a scaffold for the incorporation of new, more stable or versatile catalysts, including catalytic RNAs (ribozymes). The appearance of catalysts that could be optimized through natural selection marked a phase transition, enabling greater functional innovation.

Gradually, a diverse universe emerged, populated by increasingly complex protocells, (proto)populations, and (proto)ecosystems, along with other infrabiological or suprachemical systems, including parasitic entities. These different prebiotic entities exhibited varying capacities for interaction and evolution, resulting in a complex landscape of mutual interactions, competition, and cooperation. This dynamic interplay constituted the foundational ground for the evolution of life.

2.3 Origins of Agency

We define a *candidate for a minimal agent* as a type of system exhibiting an organization and capabilities that could plausibly evolve in the direction outlined in the previous discussion. To evaluate potential candidates, we propose the following criteria: (1) Evolvable mechanisms for action: The system must have the ability to initiate external actions with the potential for increasing complexity. Without this capacity, it cannot follow the evolutionary trajectory typically associated with the development of agency. (2) Controlled functional actions: The system must regulate its actions in a way that directly or indirectly benefits itself. This ensures that the actions contribute to the survival or adaptation of the candidate agent. If the functional effects do not benefit the agent, there would be no selective advantage to the system, limiting the evolution of more complex actions. This criterion also implies a degree of individuality in the system. We explore three potential candidates for minimal agents: *primitive protocells*, *ecological webs of protocells*, and *adaptive protocells*.

A *primitive protocell* is a self-maintaining system capable of managing energy and material flows to sustain its far-from-equilibrium organization, a requirement for its continued existence. These protocells maintain their identity with a degree of plasticity, adapting their internal processes to different environmental conditions by transitioning between stationary states. Through these self-constructing processes, they also expel metabolic waste, subtly modifying their environment over time. However, these modifications could only accumulate and have a significant impact at longer time scales, since the expelled substances would affect the environment indirectly over generations. Primitive protocells lack the ability to adaptively modify their environment in response to immediate changes (i.e., in somatic time). Their environmental impact is therefore incidental rather than functional, limiting their agency.

Within the same prebiotic scenario, a *collective ecological system* consists of metabolically complementary populations of protocells forming a primitive food web. Through their combined metabolic activities, these systems globally alter the environment, creating conditions favorable for their collective persistence. Such networks constrain abiotic cycles and other environmental parameters, thereby sustaining the ecological community over long timescales. Despite their global influence, these systems lack a clear, integrated internal organization. It remains ambiguous whether the source of functional modifications lies within the metabolic organization of individual protocells or the global interactions within the network. Moreover, their loosely integrated structure inhibits the evolution of mechanisms for sensing environmental changes or processing information to mount adaptive responses. This weak internal organization constrains their interactive and evolutionary potential.

An *adaptive protocell* differs fundamentally in its capacity to actively interact with its environment. Beyond the basic constitutive processes, these protocells possess internal *regulatory mechanisms* allowing them to respond adaptively to external changes (Fig. 2.1a). The acquisition of action-detection loops allowed the detection of environmental features and the triggering of appropriate responses—such as secretion or positional adjustments–, generating functional effects (Fig. 2.1b). This capacity for controlled interaction with the environment introduces the potential for evolutionary complexification.

Through the development of more sophisticated detectors, effectors, and intermediary systems, adaptive protocells can extend their organizational capabilities outward, functionally modifying their immediate surroundings. This ability to "export" internal organization to control the environment represents a significant

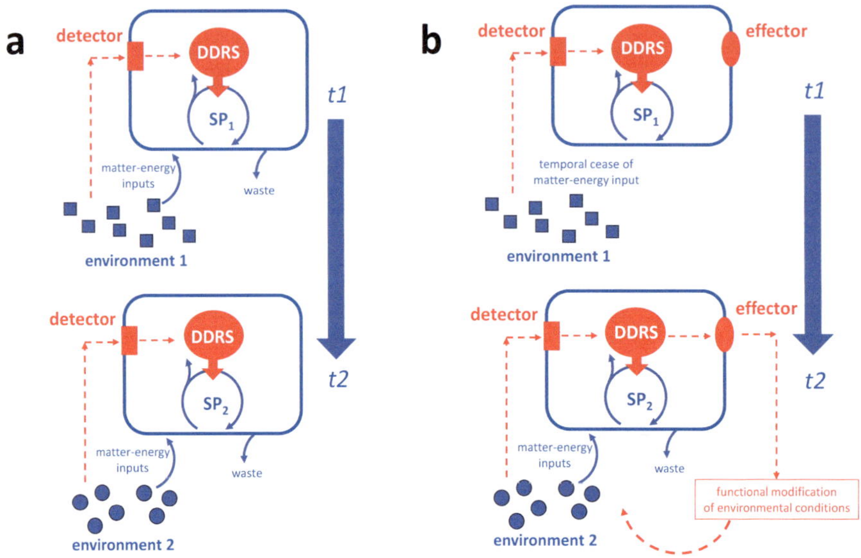

Fig. 2.1 Comparison between a merely adaptive and agential adaptive systems. (**a**) Schematic representation of an adaptive non-agential system: When the environment changes (from t1 to t2), the system can detect these changes and functionally adjust its internal organization through a regulatory subsystem to maintain its identity. (**b**) Schematic representation of an adaptive minimal agent: The system can detect relevant features of its immediate environment and initiate processes that actively modify environmental conditions when necessary (e.g., by moving). The figure illustrates the role of the regulatory subsystem, which modulates both detection and effector processes to establish successful action-detection loops. In both panels, the SP network represents the cyclic self-production processes that constitute and sustain all parts of the system, while DDRS refers to the dynamically decoupled regulatory subsystem responsible for control actions. Dashed arrows indicate relevant information flow, including environmental states detected by the system and, in panel (**b**), the system's actions on its environment. Based on Moreno (2018)

evolutionary advantage. Additionally, adaptive protocells exhibit clear individuality. Their organization becomes intrinsically linked to their capacity for action, as their survival depends on the proper functioning of membrane-bound detectors and effectors. The evolutionary improvement of these structures is interdependent with the overall system, reinforcing their status as integrated entities capable of evolutionary innovation.

Even the simplest adaptive protocells may rely on generalized solutions to recurring challenges. Many of these design patterns, such as catalytic pathways for building complex molecules or mechanisms for selective transport across cell boundaries, are ancient and widely distributed among modern life forms. These features could emerge from the intrinsic flexibility and promiscuity of primitive catalysts, such as ribozymes or early proteins. Key mechanisms enabling adaptive interaction might include: elementary feedback loops, and surface-based receptors. Based on the inherent flexibility of catalysts, feedback loops could generate oscillations and a much more complex dynamics, and they could also support primitive action-sensing cycles. Also, early protocells could develop primitive sensors on their membranes to detect environmental inputs. Simple behavioral designs (e.g., proportional, commutative, or sigmoidal responses) could enable protocells to act in response to environmental stimuli. The flexibility and degeneracy of catalysts are critical to these adaptive mechanisms. Functional degeneracy—the ability to perform multiple functions—enables a diversity of chemical reactions necessary for the protocell's constitutive processes. This flexibility also underpins basic *regulatory mechanisms*, which serve as foundational elements for interaction and adaptation.

In line with recent results obtained from computer simulations combining protocell metabolism, ecology, and evolution, we propose that the emergence of regulated interactive behavior in adaptive protocells arose from the inherent properties of their constitutive machinery. This intrinsic flexibility not only supported the diverse metabolic activities of protocells but also laid the groundwork for the development of basic regulatory and adaptive systems. These systems, in turn, enabled protocells to exert functional control over their environments, representing a critical step toward the evolution of complex life.

2.4 Lessons

The origin of agency requires the complexification of certain prebiotic systems through the development of their interactive capacities. For an evolving system to undergo such complexification in its interactions with the external world—actively driven by its internal organization—it must possess an internal structure capable of increasing in complexity. Additionally, the interactions initiated by the system must be directly or indirectly functional for each individual instance of the system during its lifetime (somatic time). Therefore, the system's interactive machinery must be organizationally integrated with its constitutive processes.

From the perspective of the biogenesis process outlined earlier, individuated protocells appear to be the most plausible candidates for this type of complexification. These protocells are intrinsically individuated, capable of self-reproduction, and serve as units of selection that encapsulate populations of diverse molecules. More fundamentally, their organizational structure is key: protocells rely on molecular components, processes, and interactions that continuously sustain one another in a tightly cyclic, self-constructing, and self-referential manner.

This organization is cohesive and highly integrated, with its various parts and processes deeply interdependent. Successful functioning requires careful coordination of distances, times, rates, and energy exchanges among these components. The pressure for such integration is inherent to any system whose identity depends on a far-from-equilibrium cyclic set of synthetic processes—a logic of self-construction that hinges on meeting specific energy demands and maintaining the precarious dynamics of its constitutive and interactive processes.

This type of organization holds significant potential for increasing complexity by incorporating additional and more diverse functional constraints. Furthermore, protocells can reproduce their organization and evolve over time. However, the increase in complexity necessitates expanding the web of endogenous (higher-order) functional constraints while ensuring their integration. This includes developing mechanisms to regulate interactions with the environment.

In other words, the potential for complexification arises from the system's ability to incorporate regulatory mechanisms that reorganize its components across differentiated levels. These mechanisms establish constraints layered on top of existing constraints, enabling protocells to achieve greater complexity and functionality while maintaining their identity.

2.5 Towards the Prokaryotic World

In this work, we aim to investigate the evolutionary origin of more complex forms of agency, tracing a continuity through their evolutionary history to identify and understand the fundamental features of agency. During biogenesis, certain systems likely thrived because their interactive capabilities enhanced their chances of survival and propagation. This makes it reasonable to hypothesize that such systems could serve as candidates for harboring the evolutionary origins of agency. If this hypothesis holds, the path to agency begins with biogenesis, contributing to the transformation of protocells into primitive organisms and progressing toward the last universal common ancestor (LUCA) and the prokaryotic forms that evolved from it (Fig. 2.2).

These primitive agents would have needed to address conflicts on at least two fronts: (1) adapting to environmental changes, whether cyclical or unforeseen, and (2) coexisting with other agents that arose from the diversification of earlier forms. To address these challenges, certain strategies would have likely emerged early in evolutionary history. For instance, primitive life cycles could have facilitated the ontogenetic deployment of diverse metabolic or structural forms capable of

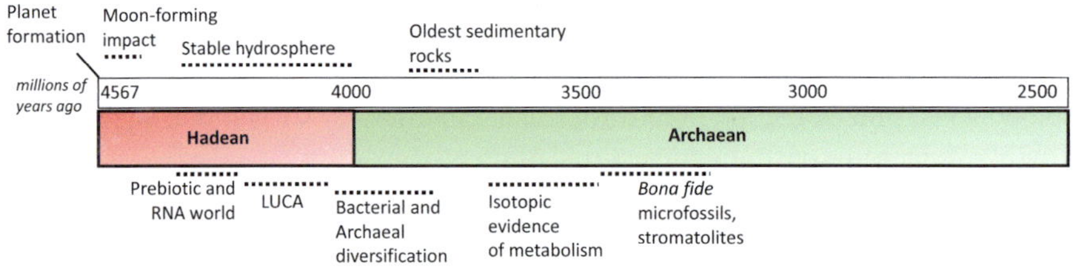

Fig. 2.2 Early history of Earth and life. Geological and planetary evidences suggest that Earth remained habitable since the establishment of a stable hydrosphere. Thus, as suggested by diverse data, including deep phylogenies, life could emerge during the Hadean. LUCA (Last Universal Common Ancestor) would be preceded by various evolutionary stages, including the so-called RNA world. On the other hand, morphological, isotopic and geochemical evidences indicate that a diversity of microbial ecosystems were active at least 3.4 billion years ago. Based on Cantine and Fournier (2018)

responding to environmental challenges. In subsequent sections, we will explore examples such as sporulation, a successful strategy used by many microorganisms to endure adverse conditions.

Agency gradually developed into more complex forms of action-detection loops, enabling organisms to respond not only to environmental changes (e.g., fluctuations in physical or chemical variables) but also to interactions with other agents—whether members of the same colony or invasive, potentially harmful entities. This interaction fostered the emergence of new domains of communication, collaboration, and defense, contributing to the increasing complexity of agency itself.

A life cycle ultimately represents a solution to a conflict that arises between reproduction and agency as complexity increases. Specifically, it involves a decoupling of two phases: (1) continuous metabolic activity associated with growth and reproduction, and (2) adaptive metabolic activities focused on detecting and managing environmental conditions, including determining the optimal timing for reproduction. This decoupling enables organisms to balance reproduction with the demands of their agential or adaptive behaviors.

In natural environments, bacteria are often found attached to surfaces, where they organize into highly structured, architecturally complex communities through the production of an extracellular matrix, collectively referred to as a biofilm. Biofilms represent a continuum of organizational complexity, which ranges from relatively simple structures to more intricate, ecologically based spatial organizations such as stromatolites—biosedimentary formations arising from the metabolic activity of spatially organized microbial communities. The emergence of these early collective organizations required simple forms of agency. These agents participated in the establishment of communities where physical proximity facilitated material exchanges, forming primitive ecosystems that resolved coexistence challenges at the early stages of microbial evolution. Unsurprisingly, stromatolites are among the oldest biological remnants found in the geological record (Fig. 2.2).

In the following chapter, we will delve into the conceptual implications of these forms of organization. The qualitative and spatial extension of the (proto)biosphere was critical for the long-term maintenance and evolution of agency, as it underpinned the establishment of biogeochemical cycles necessary for the persistence of life. The importance of favorable planetary conditions is illustrated by the hypothesis that, during Earth's early history, Mars likely had similar conditions conducive to life. However, the absence of timely biogeochemical interactions on Mars may have precluded the biological diversification necessary for sustaining life on that planet.

Bibliography

Andrews SS, Wiley HS, Sauro HM (2024) Design patterns of biological cells. BioEssays 46:2300188. https://doi.org/10.1002/bies.202300188

Bich L, Mossio M, Ruiz-Mirazo K, Moreno A (2016) Biological regulation: controlling the system from within. Biol Philos 31:237–265. https://doi.org/10.1007/s10539-015-9497-8

Cantine MD, Fournier GP (2018) Environmental adaptation from the origin of life to the last universal common ancestor. Orig Life Evol Biosph 48:35–54. https://doi.org/10.1007/s11084-017-9542-5

Carbonell P, Peretó J (2025) Before LUCA: unearthing the chemical roots of metabolism. Phil Trans R Soc B. 380:20240292. https://doi.org/10.1098/rstb.2024.0292

Di Paolo EA (2005) Autopoiesis, adaptivity, teleology, agency. Phenomenol Cogn Sci 4:429–452

Javaux EJ (2019) Challenges in evidencing the earliest traces of life. Nature 572:451–460. https://doi.org/10.1038/s41586-019-1436-4

Moreno A (2018) On minimal autonomous agency: natural and artificial. Complex Syst 27:289–313. https://doi.org/10.25088/ComplexSystems.27.3.289

Pascal R (2024) Evolutionary abilities of minimalistic physicochemical models of life processes. Chem Eur J 30:e202401780. https://doi.org/10.1002/chem.202401780

Peretó J (2019) Prebiotic chemistry that led to life. In: Kolb VM (ed) Handbook of astrobiology. CRC Press, Boca Raton, pp 219–233. https://doi.org/10.1201/b22230

Peretó J (2024) A praise of imperfection: emergence and evolution of metabolism. In: Di Mauro E (ed) The first steps of life. Wiley, Hoboken, NJ, pp 79–98. https://doi.org/10.1002/9781394264155.ch4

Ruiz-Mirazo K, Briones C, de la Escosura A (2014) Prebiotic systems chemistry: new perspectives for the origins of life. Chem Rev 114:285–366. https://doi.org/10.1021/cr2004844

Ruiz-Mirazo K, Moreno A (2024) On the evolutionary development of biological organization from complex prebiotic chemistry. In: Mossio M (ed) Organization in biology. History, philosophy and theory of the life sciences, vol 33. Springer, Cham, pp 187–218. https://doi.org/10.1007/978-3-031-38968-9_9

Shani I (2012) Setting the bar for cognitive agency: or how minimally autonomous can an autonomous agent be? New Ideas Psychol 31:151–165

Shirt-Ediss B, Ferrero-Fernández A, De Martino D et al (2024) Modelling the prebiotic origins of regulation and agency in evolving protocell ecologies. Phil Trans R Soc B 380:20240287. https://doi.org/10.1098/rstb.2024.0287

Chapter 3
A Big Bang of Agency: The Prokaryotic Way of Life

Abstract Prokaryotic organisms, encompassing bacteria and archaea, exemplify the simplest life forms and exhibit remarkable evolutionary endurance and adaptability. This chapter explores the pivotal innovations in prokaryotic agency, including secretion, motility, and the emergence of life cycles, which enhanced metabolic and genetic complexity. These mechanisms underscore the adaptability of prokaryotes, enabling them to thrive in diverse environments and contributing to their evolutionary success. Despite their impressive capabilities, prokaryotes face inherent constraints in achieving structural complexity. We conclude by contextualizing these adaptations within a broader evolutionary scenario, highlighting the transition from prokaryotic simplicity to eukaryotic complexity.

Keywords Prokaryotic agency · Motility · Secretion · Evolutionary adaptation · Life cycle

3.1 How Could Protocells Dream of Cells?

Prokaryotic cells, encompassing members of the bacterial and archaeal domains, represent the simplest morphological organization of life. Paleontological and genomic evidence strongly supports the idea that prokaryotes are the oldest living organisms (see Fig. 2.1). Consequently, it is widely accepted that the two prokaryotic domains evolved from a hypothetical last universal common ancestor (LUCA) over 3.5 billion years ago. By employing comparative genomics, scientists can infer the metabolic and genetic characteristics of LUCA, although this remains a challenging endeavor. While prokaryotes are often described in textbooks as unicellular free-living organisms, often they exhibit different forms of multicellular organization.

Á. Moreno, J. Peretó, *An Evolutionary Story of Agency*, Foundations for Interdisciplinarity in the Life Sciences: Concise Monographs,
https://doi.org/10.1007/978-3-032-05644-3_3

In the preceding chapter, we explored the hypothetical processes leading to LUCA. Although there is no consensus on the duration of these processes, it is clear that life has existed far longer than the period required for molecular complexity to progress to the emergence of LUCA. Interestingly, while all prior stages of increasing complexity before LUCA were likely erased by successive, more advanced 'proto-life' forms, prokaryotic cells have persisted throughout the subsequent history of life. This enduring presence underscores the remarkable robustness and metabolic versatility of prokaryotic organization, as well as the indispensable role of microorganisms in the long-term functioning of the biosphere. Prokaryotes are, therefore, irreplaceable.

One evolutionarily significant aspect of the prokaryotic world is the immense size of microbial populations. Prokaryotic cells outnumber unicellular eukaryotes by orders of magnitude, and unicellular eukaryotes, in turn, far exceed the populations of plants and animals. This vast population size is a crucial factor when considering the role of natural selection, genetic structure, and the effects of stochastic processes, such as genetic drift, in evolution.

What, then, distinguished the proto-organisms that preceded LUCA from the first prokaryotic cells? We can address this question by examining changes and innovations in cellular "machinery", such as the establishment of a basic but versatile metabolic network and the molecular mechanisms for accurate replication, transcription, and translation (including the universal genetic code) or the biochemical identity of prokaryotic cell membranes (Fig. 3.1). However, our focus here is on the

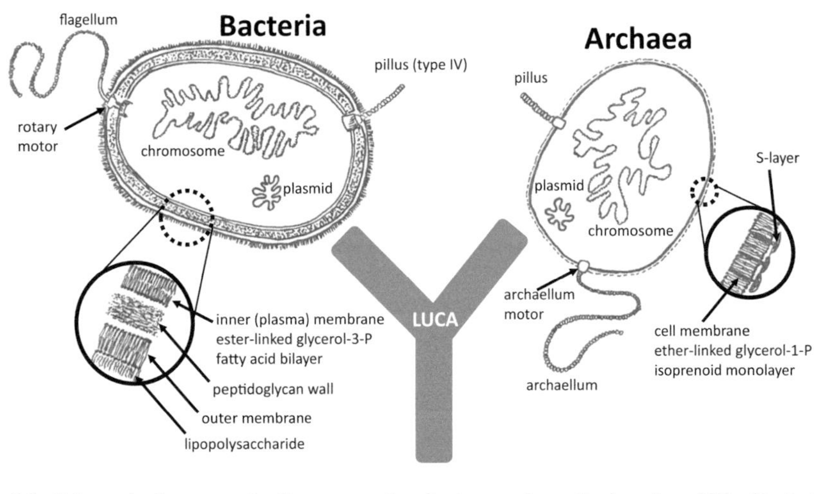

Fig. 3.1 Schematic diagrams of cells representing the two prokaryotic domains of life: Bacteria and Archaea. Both domains evolved from the last universal common ancestor (LUCA). The diagrams highlight key cellular features common to most members of each domain, with an expanded view of the cell envelope to emphasize biochemical differences between their membranes. The motility structures—the bacterial flagellum and the archaeal archaellum—have different evolutionary origins whereas pilli share a common ancestor. Adapted from Keeling (2024)

functional consequences of these developments. Specifically, we ask: what new adaptive capabilities might have arisen from the significant increase in metabolic and genetic complexity during this pivotal transition to a prokaryotic lifestyle?

"The dream of every cell is to become two cells," a phrase attributed to François Jacob by Jacques Monod, captures the centrality of reproduction to life. The uninterrupted continuity of life on Earth suggests that evolution has repeatedly found solutions to ensure the preservation of cellular reproduction despite myriad challenges. During the final stages of biogenesis before LUCA, some proto-organisms likely developed distinct, irreversible forms of organization within their life cycles. This allowed reproduction to occur across these forms, with each adapting to specific environmental conditions and possessing a unique set of adaptive strategies. Growth and reproduction would logically be tied to the form best suited to favorable conditions.

However, achieving such morphological and metabolic differentiation within a single organism's lifecycle would have posed significant challenges. Initially, the number of stages was likely minimal—perhaps only two—since these stages represented fundamentally different organizational states produced during the organism's life. Until a regulatory system capable of selectively activating or suppressing specific gene sets in response to environmental changes evolved (a feature likely present in LUCA, with its DNA genome, transcription, and translation machinery), connecting and disconnecting different functional states would have been impossible. The ability to regulate such transitions enabled organisms to adapt by modifying, replacing, or eliminating elements of their constitutive organization.

The development of different life-cycle trajectories and their control marked a profound evolutionary milestone. For the first time, organisms could exhibit a dual-level adaptive strategy: one tuned to relatively minor environmental fluctuations and another triggered by more extreme changes. Since the reproductive cycle was a central part of these transitions, the effects on fitness must have been very significant at this stage, reproductive success becoming strongly dependent on the regulation of specific life trajectories. For instance, sporulation is a tightly regulated process through which bacteria form dormant, resistant spores in response to environmental challenges such as nutrient scarcity or extreme physical and chemical conditions. This ability highlights the capacity of bacteria to sense, respond to, and adapt to their surroundings in a coordinated and goal-directed manner. Spores can endure harsh conditions and remain viable for extended periods, germinating when favorable conditions return.

Thus, the unfolding of life cycles and their control extended an organism's adaptive repertoire by aligning growth and reproduction with optimal conditions. This innovation not only enhanced the reproductive success of increasingly complex organisms, facilitating further complexity, but also bolstered their adaptivity and robustness. Life cycles, therefore, represent a strategic evolutionary development, allowing organisms to synchronize reproduction with the most favorable environmental conditions.

3.2 The Agential Dimension of the Prokaryotic Organisms: Secretion and Motility

Prokaryotes demonstrate two fundamental forms of agency: modifying their environment through chemical secretion or changing their position to find more favorable conditions via active locomotion.

In the first form, *secretory agency*, organisms alter their environmental conditions by releasing chemicals in response to external cues. This process requires detecting the environment and dynamically adjusting the type and amount of secretion to reflect real-time changes. Such selective secretion establishes a cyclic detection-action relationship, where the chemical release is causally linked to detected environmental changes, ensuring that the action effectively supports the organism's survival.

The second form, *motile agency*, involves an organism exerting force to change its spatial position using a specialized locomotion mechanism. Like secretory agency, motility requires environmental detection and a correlation between detected changes and the movements triggered. Through self-generated movement, motile agency creates adaptive detection-action cycles: movements alter environmental conditions, which in turn shape subsequent actions intended to be functional.

This interplay mirrors secretion-based agency, where modulating chemical release influences environmental factors, which then adjust future secretions. A symmetry exists between the two forms: secretion-based agency transforms spatial environmental differences into temporal changes, while motility depends on movement to perceive and respond to spatial variations, rendering it "blind" to static temporal differences.

Both forms of agency equip prokaryotes with strategies to manage external variations. While secretion-based agency is likely the most ancient form, motility also has deep evolutionary roots. Various motility mechanisms exist in bacteria and archaea (see next section for details). For example, bacteria use flagella and archaea use archaella for swimming; though functionally similar, these structures differ in composition and evolutionary origin. However, long and thin appendages, known as type IV pili (twitching motility), show clear homologies between the two domains, suggesting a shared evolutionary origin and its presence in LUCA (Fig. 3.1). Twitching motility often involves the secretion of lubricating substances that facilitate movement.

Prokaryotic cells often integrate motility and secretion, switching between or combining these strategies depending on environmental conditions. For instance, *Caulobacter crescentus* exhibits motile agency in one life cycle stage and secretory agency during its sessile phase. Motile agency is essential for navigating through physicochemical gradients, whereas secretion dominates in scenarios like pathogenic infections and biofilm formation. Both forms of agency are widespread across bacteria and archaea, which often employ similar molecular machinery for each.

Comparing their evolutionary potential, secretion-based agency appears highly versatile due to the variety of signaling substances available. However, these substances diffuse beyond the organism's control after release, limiting the ability to precisely influence their effects. In contrast, motility allows continuous control over movement direction and speed, and it can integrate multiple movement mechanisms. This flexibility expands with the evolution of sensors capable of detecting diverse inputs such as light, sound, and chemical gradients.

Motility also exerts selective pressure on sensory systems to match the demands of movement, driving the evolution of more sensitive detectors and sophisticated information processing. This dynamics creates a feedback loop of functional optimization, where increasingly complex effector-sensor systems emerge. While secretion-based agency enables modulation of substance flow and type, its potential for rapid environmental interaction is limited compared to motility. Consequently, motility has been a key driver of more advanced sensory and behavioral adaptations, making it more influential in the evolution of complex agency.

In Chap. 4, we will examine the emergence of eukaryotic agency through this lens.

3.3 The Amazing Diversity of Prokaryotic Agency

Secretory agency in prokaryotes plays a fundamental role in cell communication and collective coordination, particularly in biofilm colonies. One key example is quorum sensing, a chemical signaling mechanism that enables bacteria to detect and respond to population density. This process helps bacteria adapt to environmental changes by modulating the genetic expression of neighboring cells, thereby regulating metabolism and influencing phenotypic traits, which may be less effective at low cell densities.

Secretory agency is especially crucial in bacterial pathogenicity. Pathogenic bacteria interact with their hosts through the secretion of various substances, including toxins that damage host cells, adhesins that facilitate attachment to host surfaces, and enzymes such as proteases and glycanases that degrade extracellular matrix proteins and polysaccharides. Additionally, many pathogenic bacteria use a complex structure with an external needle (type III secretion system) to inject signaling proteins directly into host cells, a strategy that enhances bacterial infection.

Prokaryotic motility is a cornerstone of microbial agency, enabling bacteria and archaea to move toward resources, disperse, and survive in diverse environments (Fig. 3.2). Over evolutionary history, microorganisms have developed a range of motility mechanisms suited to their ecological niches, contributing to their ubiquity in habitats from deep-sea vents to human microbiomes.

Motility mechanisms can be broadly categorized into swimming in aqueous environments and movement across surfaces. Swimming is typically driven by rotating helical flagella, powered by a sophisticated rotary motor embedded in the bacterial cell wall. This motor comprises multiple proteins working cohesively, fueled by ion gradients. Advances in cryo-electron microscopy have revealed atomic-level details of the flagellar motor, including its adaptability to

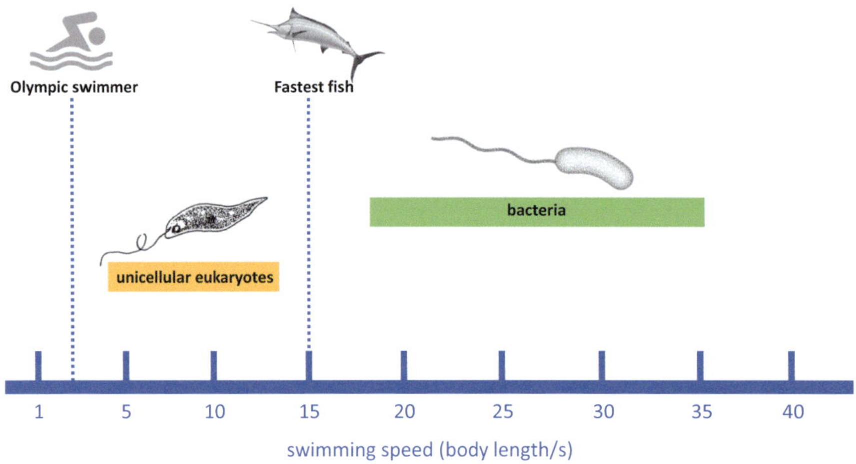

Fig. 3.2 Despite their small size and simplicity, bacteria are clear champions of the swimming world. Data from Lynch (2024)

environmental stresses through dynamic remodeling of stator units. Swimming strategies vary widely: *Escherichia coli* employs a "run-and-tumble" approach, interspersing directional runs with reorientation, while *Aliivibrio* species use a "run-reverse-flick" mechanism for rapid, precise navigation. Chemotaxis, the ability to move along chemical gradients, is pivotal in navigating complex environments (see below).

Surface-associated motility encompasses swarming, twitching, gliding, and sliding. Swarming, observed in bacteria like *Proteus* and *Aliivibrio*, involves coordinated movement on moist surfaces, often accompanied by phenotypic changes such as elongated cells and the production of many flagella. These adaptations enable bacteria to spread efficiently and form large colonies. Twitching, powered by type IV pili, relies on the rapid extension and retraction of filamentous appendages to anchor and pull cells forward, facilitating surface exploration and biofilm formation.

Gliding motility, found in groups like myxobacteria and flavobacteria, uses cell surface components that facilitate adhesion (i.e., adhesins) that move along cell-surface tracks powered by proton-driven motors. Different species employ distinct mechanisms, showcasing convergent evolution. For example, *Myxococcus xanthus* utilizes motor complexes traveling on helical tracks, while *Flavobacterium johnsoniae* recycles adhesins through a rotary system. Sliding, the simplest form of motility, involves passive outward movement driven by colony growth and reduced friction from surfactant secretion.

The diversity of forms invented by prokaryotic cells to control their spatial position goes even further. For example, in many marine planktonic prokaryotic species, there is an indirect but effective way to control their spatial position by using gas vesicles to achieve buoyancy in the water column in order to remain in the photic zone for photosynthesis.

The diversity of motility mechanisms underscores their evolutionary and ecological importance. Flagella-driven swimming excels at low Reynolds numbers (i.e. when the flow is smooth and viscous forces dominate), with adaptations like helical cell bodies enhancing propulsion. Surface motility mechanisms are advantageous for biofilm formation, host colonization, and resource acquisition. Twitching, believed to be an ancient form of motility, is widely distributed due to its simplicity.

However, motility systems also illustrate evolutionary trade-offs. Complex structures like flagellar motors and pili demand significant energy investment, balancing adaptability and resource allocation. Environmental pressures have driven convergent evolution, resulting in similar solutions across phylogenetically distant groups. For instance, the type IV pilus machinery shares structural similarities with bacterial type II secretion systems and archaeal flagella, emphasizing modular evolutionary innovation.

Despite advances, gaps remain in understanding bacterial motility, such as the molecular basis of certain gliding mechanisms and the coordination of motors in helical bacteria. Additionally, the ecological contexts of motility strategies, particularly in natural and host-associated environments, warrant further study.

3.3.1 Chemotaxis: The Precision of Chemical Navigation

Chemotaxis, the ability to navigate chemical gradients, is a fundamental aspect of microbial agency, shaping bacterial ecology, evolution, and interactions. It enables bacteria to move toward beneficial chemicals or away from harmful substances, enhancing survival and fitness. The classical model, exemplified by *E. coli*, describes a "run-and-tumble" strategy modulated by chemoreceptors that detect temporal changes in chemical concentrations. Signal transduction pathways translate external chemical cues into directional motility by controlling flagellar rotation.

Traditional models posited that chemotaxis is ineffective for small-scale interactions, as gradients from targets smaller than 4 μm were considered undetectable. However, recent evidence challenges this view, revealing that some bacteria possess exceptional chemotactic precision, enabling responses to microscale gradients produced by smaller bacterial targets. These findings suggest a broader ecological role for chemotaxis in microbial interactions.

Recent studies show that chemotaxis mediates metabolite exchange among bacteria. Cells release metabolites through exudation or lysis, forming gradients exploitable by chemotactic neighbors. Such sensing fosters cooperative or competitive relationships. For example, heterotrophic bacteria like *Marinobacter adhaerens* exhibit chemotaxis toward photosynthetic *Synechococcus*, facilitating nutrient exchange. These interactions are transient, lasting seconds, contrasting with the prolonged associations of eukaryotic hosts.

In aquatic environments, chemotaxis promotes clustering around resource patches, amplifying metabolite flux and biogeochemical activity. This is particularly important in nutrient-poor systems like the open ocean, dominated by small photosynthetic bacteria. The precision of bacterial chemotaxis is influenced by factors such as swimming speed, gradient sensitivity, and signaling efficiency. High-speed swimmers like *Aliivibrio ordalii* excel in weak gradients but risk increased diffusivity, which can reduce gradient retention. Some species offset this by adjusting turning frequency during rapid movement. The diversity of chemotactic responses reflects the variety of chemical gradients bacteria encounter. Small, diffusing molecules attract bacteria from afar, while larger compounds generate localized gradients favoring selective interactions.

Beyond nutrient acquisition, chemotaxis facilitates consortia formation, as seen in motile sulfur-reducing bacteria forming associations with phototrophic consortia in lakes. It also enhances predator-prey dynamics, aiding bacterial predators like *Bdellovibrio bacteriovorus* in locating prey. Additionally, chemotaxis may boost horizontal gene transfer by increasing physical contact between cells, influencing microbial evolution.

In conclusion, bacterial motility and chemotaxis exemplify the interplay of evolutionary innovation, biophysical constraints, and ecological adaptability. By navigating physical and chemical landscapes with precision, bacterial agents not only optimize survival but also shape microbial communities and ecosystems.

3.4 Building Collective Forms of Agency

The emergence of collective behaviors in bacteria underscores their complexity and adaptability. These behaviors arise from interactions between individual bacterial cells and their environment, leading to coordinated actions that enhance the survival, efficiency, and resilience of bacterial populations. Key processes driving such behaviors include quorum sensing and biofilm formation.

As mentioned before, quorum sensing is a cell-density-dependent communication mechanism where bacteria produce, release, and detect signaling molecules called autoinducers. As the bacterial population increases, the concentration of autoinducers rises. Once a threshold concentration is reached, it triggers a synchronized response across the bacterial community. A classic example is the bioluminescence of *Aliivibrio fischeri*, a marine bacterium that emits light at high cell densities, benefiting its symbiotic relationship with certain marine animals.

Bacterial biofilm formation is a highly organized, multi-step process that exemplifies biological agency. Biofilms are structured communities of bacteria that adhere to surfaces and are embedded in a self-produced extracellular matrix. This matrix, composed of polysaccharides, proteins, and nucleic acids, protects bacteria from environmental stresses, antibiotics, and host immune responses, enhancing their survival and persistence. Biofilm formation involves environmental sensing, coordinated behavior, and adaptation to dynamic conditions—hallmarks of biological agency.

Certain bacteria, such as myxobacteria and cyanobacteria, exhibit remarkable collective behaviors that highlight their capacity for cellular cooperation and coordination. Myxobacteria, a group of soil-dwelling bacteria, display some of the most complex collective behaviors observed in prokaryotes. These include social motility, predation and aggregation leading to a multicellular stage, i.e., fruiting body formation and sporulation.

Myxobacteria can move collectively as coordinated groups, forming rippling patterns or swarms. For example, *Myxococcus xanthus* employs two types of motilities—gliding, powered by the secretion of polysaccharide slime, and social motility, driven by the action of type IV pili. These pili enable synchronized group movements. *M. xanthus* preys on other microorganisms by secreting lytic enzymes and specialized metabolites that break down prey cells, allowing nutrient acquisition. However, under nutrient-limited conditions, *M. xanthus* cells aggregate to form multicellular, spore-bearing fruiting bodies. This process involves several stages, including aggregation, mound formation, and differentiation into stress-resistant spores. Fruiting bodies enable the bacteria to survive desiccation, UV radiation, and other environmental stresses. Myxobacteria rely on quorum sensing and additional signaling mechanisms to coordinate these behaviors. Environmental cues, such as nutrient levels and stress conditions, regulate their transition between growth and development modes.

Worth commenting, as well, cyanobacteria constitute a diverse group of photosynthetic bacteria that exhibit a range of collective behaviors, enabling them to adapt to diverse environmental conditions. These behaviors play critical roles in ecological processes, such as nitrogen fixation and primary production in aquatic ecosystems. Cyanobacteria often grow in colonies or filaments, with cells adhering to each other via extracellular polysaccharides and matrix components. Filamentous cyanobacteria, such as *Anabaena*, can differentiate into specialized cells which fix nitrogen under nitrogen-limited conditions. Some cyanobacteria, such as *Oscillatoria*, exhibit gliding motility and phototaxis, moving toward or away from light sources. This movement is often driven by type IV pili and polysaccharide secretion.

Cyanobacteria use quorum sensing to regulate collective behaviors, such as biofilm formation and stress responses. For instance, *Synechocystis* employs quorum sensing to coordinate gene expression in response to cell density. Cyanobacteria also form biofilms that enhance their protection and resource utilization. These biofilms, composed of extracellular polymeric substances, allow cyanobacteria to adhere to surfaces and contribute to ecological processes. For example, *Nostoc* species form biofilms in soil and freshwater environments, playing a role in nutrient cycling and soil stabilization.

The collective behaviors of bacteria demonstrate their remarkable adaptability and evolutionary success. These behaviors illustrate how bacterial populations use coordinated strategies to optimize survival, respond to environmental changes, and thrive in diverse ecosystems. Ultimately, collective coordination ensures a strong connection between life cycles and agency, allowing bacteria to function as highly adaptive and resilient communities.

3.5 Towards the Eukaryotic Complexity

As we have seen, prokaryotic organisms exhibit an extraordinary diversity of life forms. They have adapted to extreme environments, developed various modes of interaction, and existed in both unicellular and multicellular states. Many have evolved life cycles that include dormant phases or multicellular stages, displaying collective forms of agency. However, their potential for increasing in size—and consequently, for greater organismal complexity—faces inherent limitations.

In collective systems like biofilms, the lack of sufficient organismal integration prevents them from becoming fully multicellular entities. Similarly, in associative microorganisms, the transition to a multicellular state remains optional and is triggered by environmental cues rather than being a stable developmental feature. As a result, despite their extensive evolutionary history and remarkable adaptability, prokaryotes have not achieved the structural, developmental, or agential complexity seen in other life forms. This raises a fundamental question: why?

Stephen J. Gould argued that while the biosphere is predominantly populated by bacteria, situated at the lower end of the complexity spectrum, the evolutionary pathway to greater complexity requires a fundamental transformation of prokaryotic microorganisms. Richard E. Michod further suggested that the evolution of higher complexity depends on resolving evolutionary conflicts at lower organizational levels. For larger systems to function as cohesive units capable of complex behaviors, their components must integrate and cooperate effectively. Prokaryotic limited exploration of multicellularity, as well as their relatively simple achievements on that front, illustrate the inherent constraints on their complexity.

Sergio Muñoz-Gómez has emphasized that cellular complexity comes with significant energetic costs. In the short term, new genes with weak or unoptimized functions demand more energy, as they occupy proteome resources without immediately enhancing fitness. In the long term, the accumulation of many new genes strains the proteome further, diverting resources away from growth and reproduction. While these additions do not necessarily increase absolute energy demands, they reallocate energy from essential cellular processes to maintaining these new features. This perspective aligns with phylogenetic scaling laws, which show a decrease in ribosome concentration and productivity as cell size and complexity increase. Consequently, a substantial rise in cellular complexity places organisms in a distinct adaptive zone where energy is primarily allocated to survival rather than reproduction. This shift helps explain why prokaryotes have not overcome their "complexity ceiling" through simple associations of similar single-celled entities but instead through the evolution of larger, more integrated cells.

Achieving such a breakthrough has been an exceptionally challenging process, marked by conflicts and requiring radical cellular reorganization. This transformation involved extensive genetic acquisitions and losses, as well as endosymbiotic events. The intimate symbiosis between different prokaryotic species inherently created conflicts, and the historical resolutions to these conflicts were both rare and unique. Notably, eukaryotes, the outcome of this process, share a common ancestor (are monophyletic), and there are no known examples of free-living prokaryotic endosymbiotic consortia in the natural world.

In the next chapter, we will delve into the unique sequence of events that led to the origin of the eukaryotic cell, a process that represents one of the most significant transitions in the history of life.

Bibliography

Arnellos A, Moreno A (2015) Multicellular agency: an organizational view. Biol Philos 30:333–357. https://doi.org/10.1007/s10539-015-9484-0

Arnellos A, Moreno A, Ruiz-Mirazo K (2014) Organizational requirements for multicellular autonomy: insights from a comparative case study. Biol Philos 29:851–884. https://doi.org/10.1007/s10539-013-9387-x

Barandiaran X, Moreno A (2009) Adaptivity: from metabolism to behavior. Adapt Behav 16:325–344. https://doi.org/10.1177/1059712308093868

Barandiaran X (2008) Mental life: a naturalized approach to the autonomy of cognitive agents. PhD Thesis, University of the Basque Country

Berleman J, Kirby J (2009) Deciphering the hunting strategy of a bacterial wolfpack. FEMS Microbiol Rev 33:942–957. https://doi.org/10.1111/j.1574-6976.2009.00185.x

Gould SJ (1994) The evolution of life on the Earth. Sci Am 271:84–91. https://www.jstor.org/stable/24942873

Keeling PJ (2024) Horizontal gene transfer in eukaryotes: aligning theory with data. Nat Rev Genet 25:416–430. https://doi.org/10.1038/s41576-023-00688-5

Lynch MR (2024) Evolutionary cell biology. The origin of cellular architecture. Oxford University Press, Oxford

Michod R, Roze D (2001) Cooperation and conflict in the evolution of multicellularity. Heredity 86:1–7. https://doi.org/10.1046/j.1365-2540.2001.00808.x

Miyata M, Robinson RC, Uyeda TQ et al (2020) Tree of motility. a proposed history of motility systems in the tree of life. Genes Cells 25:6–21. https://doi.org/10.1111/gtc.12737

Moreno A, Mossio M (2015) Biological autonomy: a philosophical and theoretical enquiry. Springer, Dordrecht. https://doi.org/10.1007/978-94-017-9837-2

Muñoz-Gómez S (2024) The energetic costs of cellular complexity in evolution. Trends Microbiol 32:746–755. https://doi.org/10.1016/j.tim.2024.01.003

Shapiro JA (1998) Thinking about bacterial populations as multicellular organisms. Annu Rev Microbiol 52:81–104. https://doi.org/10.1146/annurev.micro.52.1.81

van Duijn M, Keijzer F, Franken D (2006) Principles of minimal cognition: casting cognition as sensorimotor coordination. Adapt Behav 14:157–170. https://doi.org/10.1177/105971230601400207

Chapter 4
Eukaryotic Agency: Looking for the Path of Agency Complexification

Abstract This chapter explores the evolutionary complexification of eukaryotic agency, emphasizing the remarkable behavioral and structural innovations in protists. Protists leveraged mechanisms developed during eukaryogenesis to exploit their larger cell size, leading to advanced motility, sensory systems, and environmental adaptability. These innovations enabled them to diversify and thrive across a wide range of ecological niches, innovations that in some branches have developed early complex forms of agency. The chapter also examines the limitations and potential of multicellular protists, highlighting how motility-based agency has driven greater complexification compared to secretion-based strategies. Insights into the transition toward multicellularity provide a foundation for understanding the evolution of further complex manifestations of agency.

Keywords Eukaryogenesis · Protist agency · Motility · Simple multicellularity

4.1 Eukaryogenesis or How Prokaryotes Broke Through the Complexity Ceiling

The terms "prokaryote" and "eukaryote" are useful for studying life from a cellular perspective, but they do not accurately reflect phylogenetic relationships. The chimeric nature of eukaryotic genomes—featuring genes of bacterial, archaeal, and viral origins—supports the long-standing hypothesis of a symbiotic origin for eukaryotes and highlights the significance of horizontal gene transfer. Phylogenetic analyses suggest that eukaryotes evolved from archaea, attaining a certain level of cellular complexity before acquiring intracellular bacteria (endosymbiosis) to form mitochondria. However, the exact timeline of these events remains uncertain. Cellular complexity may have started around 3 billion years ago (a billion years after LUCA), with mitochondrial incorporation occurring roughly a billion years

Á. Moreno, J. Peretó, *An Evolutionary Story of Agency*, Foundations for Interdisciplinarity in the Life Sciences: Concise Monographs,
https://doi.org/10.1007/978-3-032-05644-3_4

later. Despite these uncertainties, it is widely accepted that eukaryotic cells originated only once in Earth's history. Following the acquisition of mitochondria, LECA (the Last Eukaryotic Common Ancestor) emerged as a distinct entity with cellular and genomic characteristics unique to eukaryotes. This innovation was followed by a rapid diversification that produced the major eukaryotic lineages.

This diversification likely stemmed from genetic mechanisms like genetic recombination and meiotic sex, which is absent in prokaryotes, and from extensive gene duplication triggered by the transfer of genes from the protomitochondrion to the nuclear genome. Several models have been proposed to explain the chimeric nuclear genome of eukaryotes and the monophyletic origin of mitochondria. Although many details of eukaryogenesis remain unresolved, key ideas can be summarized. An ancestral archaeon, related to modern Asgard archaea, developed features such as membrane flexibility, vesicle formation, phagocytosis, and cytoskeletal elements. Notably, the genome of some Asgard clades contain many molecular eukaryotic signatures, including actin families, and the cultured Asgard archaeon *Promethearchaeum syntrophicum* exhibits symbiotic interactions and dynamic membrane structures, hinting that ancestral archaea involved in eukaryogenesis might have maintained similar traits.

The First Eukaryotic Common Ancestor (FECA) acquired additional traits through lateral gene transfer from diverse bacterial and viral sources, possibly leading to genome compartmentalization in a primitive nucleus. Over time, permanent endosymbiosis with a bacterium related to modern alpha-proteobacteria resulted in the mitochondria. This event was marked by rapid endosymbiotic gene transfer, during which many protomitochondrial genes migrated to the nucleus while others were lost. Despite the accumulation of non-coding DNA, introns, and transposons, the spatial separation of transcription and translation, particularly the maturation of mRNA by the spliceosome, gave nucleated cells a significant evolutionary advantage.

Comparative studies of eukaryotic complexity suggest that the Last Eukaryotic Common Ancestor (LECA) was a flagellated heterotroph capable of phagocytosis. LECA possessed mitochondria, an endomembrane system, a cytoskeleton composed of actin and tubulin, and a genome architecture distinct from that of prokaryotes (Fig. 4.1). During the transition from the First Eukaryotic Common Ancestor (FECA) to LECA, molecular motors such as dynein, kinesin, and myosin diversified through gene duplication and divergence. In combination with the actin-based cytoskeleton, these motor proteins provided a versatile and powerful mechanism not only for targeted intracellular transport but also for rapid and reversible changes in cell shape and various forms of motility.

A key innovation in eukaryogenesis was the emergence of actomyosin networks. In eukaryotic cells, actin filaments and myosin proteins form highly organized, dynamic structures known as actomyosin networks. These networks actively regulate cell shape and internal spatial organization (e.g., intracellular transport), playing a central role in the structure and function of much larger and more complex eukaryotic cells. Crucially, they enabled the evolution of new forms of motility. Actomyosin networks can undergo both reversible and irreversible changes:

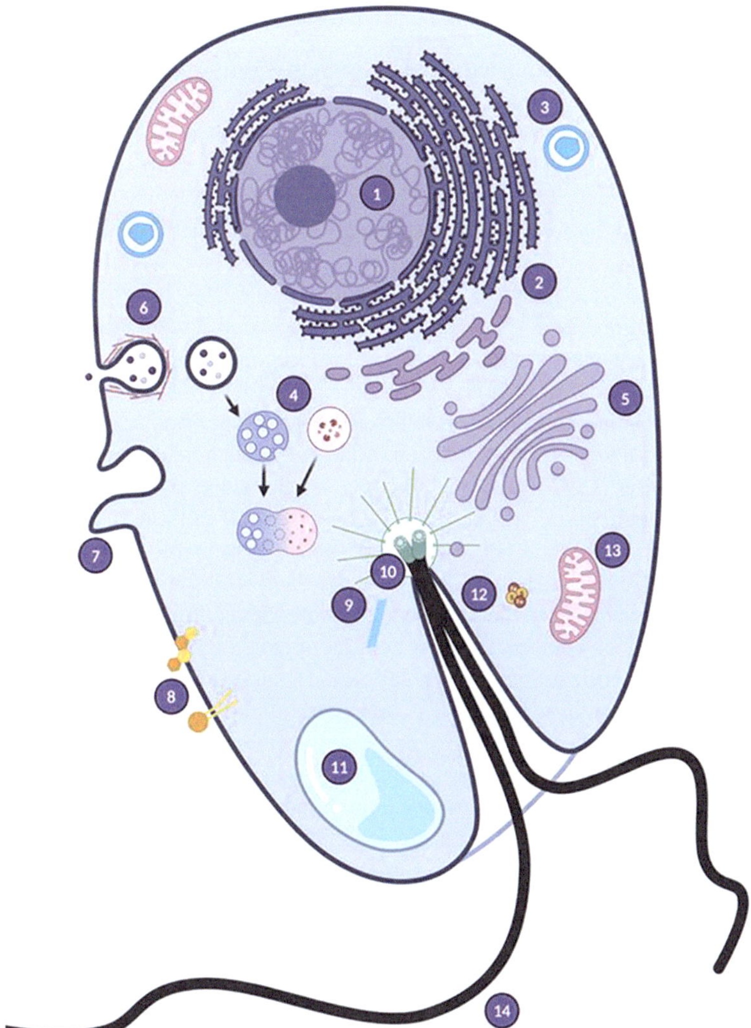

Fig. 4.1 The reconstruction of the last eukaryotic common ancestor (LECA) highlight its complex cellular and molecular biology. LECA likely possessed many features aligned with modern heterotrophic flagellated protists: (1) Nucleus; nuclear pore; linear chromosomes; spliceosome and intron-containing transcripts; meiosis, mitosis, complex cell cycle. (2) Endoplasmic reticulum. (3) Peroxisome. (4) Endo/lysosome. (5) Golgi complex. (6) End/exocytosis, actin-based. (7) Pseudopodium. (8) Membrane of glycerol-3-P, ester-linked fatty acid phospholipids and sterols. (9) Microtubule-based cytoskeleton and organizing center. (10) Basal bodies. (11) Vacuoles. (12) Cytosolic iron-sulfur cluster assembly (CIA) system. (13) Mitochondria. (14) Flagella (microtubule based). Fragment of Fig. 2 from Richards TA et al. (2024) reproduced from PLoS Biology under Creative Commons Attribution License. https://doi.org/10.1371/journal.pbio.3002917.g002

developmental processes typically rely on irreversible modifications, while behavioral responses depend on reversible ones.

Unlike prokaryotic cells, in which cytokinesis and motility are distinct processes, eukaryotic cells leverage the dynamic plasticity of the actomyosin system for a broad range of functions. This includes not only cytokinesis during the mitotic cell division, but also internal and external contractility, along with various modes of cellular locomotion.

Eukaryotes also evolved meiosis, allowing for recombination between homologous chromosomes and promoting genetic diversity. The advent of meiotic sex enabled haploid cells to fuse and complete full sexual cycles, fostering further evolutionary innovation. The alternation between asexual reproduction and sexual cycles gave eukaryotes the ability to balance rapid population growth with genetic experimentation and adaptability.

The origin of mitochondria is often linked to bioenergetic advantages, but many eukaryotic traits likely emerged before mitochondria became the center of cellular energy production. LECA's unique features set the stage for the "big bang" of eukaryotic diversification, which occurred over 10–100 million years. This rapid radiation was likely driven by genetic isolation mechanisms, including massive gene migration from mitochondria to the nucleus and widespread gene duplication, both of which are central to the emergence of biological complexity.

The emergence of a completely new cellular organization during eukaryogenesis marked, also, a significant increase in cell size. Eukaryotic cells owe their size and shape to internal protein scaffolds (e.g., actin and tubulin-based cytoskeletons) and, in some cases, external structures like cell walls. As mentioned before, the cytoskeleton likely emerged from the Asgard-related archaeal host, and it was already established in LECA, as revealed by metagenomic and biochemical studies of modern Asgard archaea. Although cell walls are found in both prokaryotes and eukaryotes, contributing to osmotic regulation and structural support, their biochemical composition is neatly different: bacterial walls are based on peptidoglycan, whereas eukaryotic walls rely on other materials, like chitin or cellulose. The independent evolution of eukaryotic cell walls highlights how different biophysical challenges can be addressed with distinct molecular solutions.

4.2 Eukaryogenesis and Agency

The newly evolved molecular machinery allowed eukaryotic cells to take advantage of a larger body size compared to their smaller prokaryotic counterparts. This innovation enabled them not only to develop efficient and rapid motility—despite the significant energetic and organizational costs of moving larger bodies—but also to generate a wide range of novel behaviors. These enhanced interactive capacities, in turn, played a crucial role in their evolutionary success.

As mentioned earlier, early eukaryotic cells exhibited an extraordinary capacity for rapid diversification. Protists—eukaryotes that are not fungi, plants, or animals, most of them free-living unicellular species—quickly conquered new niches in almost every environment on the planet. They adapted in various ways to survive and thrive under diverse conditions. Beyond their remarkable ability to form new symbiotic associations, this success cannot be fully explained without considering the extraordinary innovations of protist agency.

How and when did these innovations occur? To what extent did the early eukaryotic design influence the interactive lifestyles of protist ancestors, ensuring their success in exploiting new niches? Initially, changes in interactive capabilities may have been gradual, as there was not always a direct relationship between internal changes and their external manifestations. Nevertheless, internal changes fundamentally influenced how these new cells interacted with their environment.

Prokaryotic cells primarily rely on cytoplasmic diffusion for intracellular transport. However, as cell size increased during eukaryogenesis, machine-assisted cytoplasmic transport became essential. Several cytoplasmic changes occurred, including the organization of a dynamic cytoskeleton with protein motors and vesicular systems. Relatively advanced actin networks likely evolved prior to eukaryotic diversification. This reflects the limited diversity of regulators and network architectures in bacteria and archaea and underscores the evolutionary leap within eukaryotes. This leap included the emergence, amplification, and diversification of molecular motors during eukaryogenesis.

A dynamic cytoskeleton, enriched with numerous actin-related and actin-binding proteins, played a crucial role in the emergence, evolution, and diversification of Asgard archaea. The extensive surface area generated by a complex network of cellular protrusions, along with an unusual cell envelope, may have facilitated the intricate cell–cell interactions necessary for eukaryogenesis. Given the lifestyles of cultured Asgard strains, these interactions likely involved interspecies dependencies in syntrophic relationships, potentially leading to phagotrophic behaviors and extensive lateral gene transfer. These findings support the hypothesis that mitochondrial acquisition occurred gradually through protrusion-mediated cell-cell interactions. Therefore, the ability of ancient Asgard archaea to exhibit phagotrophic behavior may have been a key factor in the complex evolutionary transition that led to the first eukaryotic cells.

That new cellular organization radically transformed developmental and physiological mechanisms, as well as how cells interacted with their environment. New forms of sensing and responding to stimuli evolved, allowing deterministic tracking of gradients and movement along vectorial cues in three-dimensional environments (3D taxis). Eukaryotes' rich behavioral repertoire emerged thanks to several cellular innovations, including an expanded repertoire of ion channels, the emergence of cilia and pseudopodia, endomembrane trafficking, a flexible plasma membrane, and the relocation of chemiosmotic ATP synthesis from the plasma membrane to mitochondria. This shift liberated the plasma membrane for more complex electrical signaling involved in sensing and responding to environmental stimuli. As cell size increased, these innovations collectively allowed eukaryotes to surpass prokaryotes in movement precision and speed, as seen in chemotaxis, active food selection, phagotrophy, and regulated cell–cell fusion during sexual reproduction. These advances in behavior and sensing were among the key contributors to the success of eukaryotes.

The emergence of unicellular eukaryotic organisms illustrates how secretion and motility define distinct evolutionary paths toward the increasing complexity of agency, as discussed in the next section. One of the most critical agential innovations in early eukaryotes was related to feeding by phagotrophy. As mentioned earlier, LECA already was a heterotrophic flagellate with a sophisticated cytoskeletal architecture that enabled phagocytosis. Amoebae, for instance, can alter their shape by creating temporary cytoplasmic extensions known as pseudopodia, used for both feeding and movement. Although some argue that phagocytosis is energetically favorable only in the presence of mitochondria and requires the full complexity of a eukaryotic cell, amitochondriate eukaryotes are also capable of phagocytosis, as probably also were the ancestral archaea before the acquisition of the primitive mitochondrion.

Actually, primitive eukaryotic cells are bound to have developed new feeding strategies, notably more sophisticated predation behaviors. Early eukaryotes may have actively explored and selected particles during feeding. If so, the cytoskeleton and endomembrane system are key to understanding the rich behavioral repertoire of protists. These structures support the formation of cilia and flagella during the cell cycle. The cytoskeleton not only organizes cell shape but also enables locomotion and organelle movement, and supports various sensory and motility functions as we will discuss in the following sections.

4.3 The Remarkable Diversity of Unicellular Protist Agency

Closely linked to the internal innovations that paved the way to eukaryogenesis, protists introduced a wide and rich array of new mechanisms that served to expand secretory and motile agency in the biological world. As previously noted, these organizational innovations have enabled eukaryotes to occupy a wide range of new ecological niches. In turn, adaptation to these diverse environments has shaped both the type and degree of agency complexity across different evolutionary lineages. For example, many planktonic eukaryotes have entirely lost motility, yet this has not diminished their evolutionary success or internal complexity. In the following sections, we will focus on those evolutionary pathways that have led to more complex forms of agency and explore how these new capabilities are integrated into and supported by changes in internal organization.

4.3.1 Secretion

Secretion-based agency is a universally used form of adaptive interaction, which both non-motile and motile protists have developed. Protists exhibit novel ways to control their secretory responses. One prominent example is secretion via vesicles (exocytosis), a widely utilized mechanism in protists. This ability relies on intracellular membranous structures, known as secretory organelles, which are equipped with molecular machinery to accumulate and store various types of cargo. Remarkably, protists can also secrete entire packages of chemicals—or even organelles—offering a far more precise and effective means of action compared to the diffusion-based chemical release typical of prokaryotes.

An even more intriguing innovation, particularly related to predation in motile protists, is the development of specialized structures called extrusomes. These organelles enable protists to engage in a wide range of interactions, combining secretion and motility, as the action involves ejecting a targeted structure, such as a mucous mass (mucocyst), a filament (trichocyst), or a structure that releases toxic components (toxicyst) upon reaching its target.

4.3.2 Motility

When analyzing eukaryotic agency, the most fascinating innovations are found in motile agency. This stems largely from the specific size of eukaryotic cells, which profoundly influences how they move. Eukaryotic cells are significantly larger than prokaryotes, and this obliges them to navigate a world where viscous forces dominate over inertia. At this scale, movement is governed by a complex interplay of physical forces and mechanical constraints, prompting protists to develop advanced propulsion and navigation strategies. These innovations have, in turn, greatly expanded their behavioral repertoire.

The differences in size also influence movement in other ways. For instance, the larger size of eukaryotic cells allows them to detect gradient differences while swimming by comparing the asymmetric activation of receptors distributed across their membrane. This capability enables them to direct their movement toward a target or away from harmful stimuli. Additionally, their locomotion is organized differently than in prokaryotes, relying on the synchronized activity of molecular motors operating at microtubules or actin filaments.

Eukaryotic motility exhibits an incredible diversity of behaviors, encompassing strategies such as swimming, gliding, ejecting toxic or needle-like structures, contracting their shape to feed, and engulfing prey. Some protists can rapidly change direction, while others actively direct water flows through cilia toward an oral cavity for feeding. Slower-moving protists utilize specialized microtubule-filled appendages—such as axopodia, haptonema, and other tentacle-like structures—for interception feeding, prey capture, and transport. As mentioned before, many protists have organelles that allow them certain rapid reaction behaviors. Although these reactions differ from whole-organism motility, they are often synchronized with the detection of distant stimuli. This type of agency requires a sensor-effector coupling akin to that of fast motile agency and may even be better categorized as motile rather than strictly secretory.

Eukaryotic motility primarily relies on flagella (also known as cilia or undulipodia), which are structurally and biochemically distinct from bacterial flagella and archaeal archaella. Thus, the eukaryotic cilium differs fundamentally from the prokaryotic flagellum in its molecular structure and bioenergetic mechanisms. Eukaryotic cilia operate with their own intricate mechanisms of movement, supported by a core structure (or axoneme) with nine peripheral microtubule doublets encircling a central pair (9+2 axoneme). This structure forms a scaffold for the assembly of inner and outer dynein arms. These arms enable movement through dynein motor activity. Protist cilia are considered organelles in their own right, equipped with a localized "metabolism" to sustain their functions.

Powered by ATP-dependent molecular motors and cytoskeletal components, flagella are nearly ubiquitous in eukaryotes, suggesting their presence in LECA. During the transition from FECA to LECA, primitive eukaryotes developed polarity as microtubules grew asymmetrically from precursors to the microtubule-organizing center. This polarized morphology created differentiated membrane domains with specific receptors for environmental sensing and signal transduction. Vesicle trafficking maintained this polarization and enabled gliding motility. The addition of molecular motors like dynein transformed this system into a flagellum with a conserved 9+2 microtubule axoneme structure.

Intraflagellar transport (IFT) became essential for assembling and maintaining flagella, which are differentiated compartments with specialized lipid and protein compositions. IFT also supported secretion, as vesicle fusion at the membrane released vesicular contents externally. This capability underscores how eukaryotic cells express secretion-dependent functions in ways fundamentally different from prokaryotes.

This evolutionary perspective highlights the role of cellular asymmetry and intracellular transport in the development of flagella. A polarized cell surface with specialized receptors facilitated environmental sensing and motility, likely forming the ancestral state of the modern eukaryotic axoneme. By integrating sensory and motility functions, primitive flagella would have increased receptor exposure, setting the stage for the extraordinary diversity and wealth of eukaryotic motility mechanisms observed today.

4.4 How Unicellular Protists Explore the World

4.4.1 Complexification of Behavior in Protists

Unicellular protists exhibit a diverse array of motility mechanisms, including whole-body locomotion via cilia and a highly adaptable form of movement through pseudopodia, commonly referred to as "amoeboid motility." This type of movement, characterized by dynamic shape changes, is unique to eukaryotes and absent in prokaryotes.

The remarkable diversity of protist motility behaviors is underpinned by a sophisticated array of molecular mechanisms, as discussed above. Efficient internal transport systems employs motor proteins like myosin, kinesin, and dynein, which move along actin filaments and microtubules. However, it is the dynamic and adaptable nature of the cytoskeleton that truly enables their versatile motile capacities. Dynein's interaction with microtubules drives the motion of eukaryotic flagella and cilia, while amoeboid motility relies on the flexible contraction of bipolar myosin and actin filaments. The tradeoff for the amoeboid movement's plasticity is keeping a relatively low speed, as the larger cellular volume and reliance on slow diffusion processes complicate the integration of sensorimotor pathways.

Protist movements are often initiated by mechanical stimuli and involve calcium-dependent changes in microtubule configurations. In larger protists, motor actions are frequently mediated through the membrane via changes in electric potential. These rapid responses typically involve unidirectional electrical pulses, characterized by fast, regenerative shifts in membrane potential.

Ciliary locomotion provides protists with their fastest swimming movements, enabling speeds between 400 and 2000 μm/s. For instance, *Paramecium* has approximately 4000 motile cilia arranged in longitudinal rows, which move synchronously, driven by a calcium-based action potential. These cilia enable *Paramecium* to navigate helical trajectories punctuated by sudden directional shifts called avoiding reactions. These reactions, triggered by unfavorable conditions such as obstacles or harmful chemicals, form the basis of the organism's "trial-and-error" behavioral strategy.

Another example of motile complexity in protists is found in dinoflagellates, which possess two perpendicular flagella: a transverse flagellum and a longitudinal trailing flagellum. The transverse flagellum induces rotation around the cell's longitudinal axis, while the trailing flagellum contributes to both forward translation and rotation around a perpendicular axis. These combined movements produce a helical swimming trajectory. Dinoflagellates can independently adjust the rotational components and translational velocity to adapt to their environment. During hunting, many dinoflagellates use stinging darts to capture a protist prey—as seen in species of the genera *Polykrikos*–, detecting vibrations when the prey approaches. However, they often discharge multiple darts before successfully striking a target.

4.4.2 Complexification of Protists Sense Organs

Protists possess a diverse range of sensory structures capable of detecting a wide variety of environmental inputs, including light, sound, chemicals, gravity, motion, and temperature. These stimuli are processed differently depending on the protist species, enabling them to perform behaviors such as moving toward the water surface through gravitaxis, avoiding obstacles using mechanosensitivity, resisting water currents through rheotaxis, and avoiding bright light or extreme temperatures. While many of these capabilities are also found in prokaryotes, the way protists process and integrate sensory signals—and the speed and mechanics of their responses—are fundamentally different.

One of the most significant evolutionary innovations in protists is their ability to respond to light stimuli. For the first time in evolutionary history, eukaryotes developed the ability to follow light direction in three dimensions under open-water conditions. This phototactic ability is mediated by eyespots, which detect light intensity and direction, enabling protists to either swim toward or away from light, depending on environmental conditions. A notable example is the unicellular algae *Euglena*. Its eyespot senses light intensity and directs the cell toward brighter areas. This fast, precise swimming behavior is enabled by an extremely elastic motor system that adjusts the organism's shape and movement dynamically. By coordinating eyespot signals, cell orientation, and light detection, *Euglena* can navigate structured light fields, avoid edges, and follow light gradients effectively.

Even more remarkable are the ocelloids found in warnowiid dinoflagellates, slow-moving protists that hunt prey by ejecting needle-like darts. Ocelloids are among the most complex subcellular sensory structures known. They consist of a lens for concentrating light, surrounded by mitochondria functioning as a cornea, and a pigmented retinal body with intricate membrane structures. These components work together to detect distant movements and trigger a protist's extrusive, harpoon-like structure against prey. It is hypothesized that ocelloids evolved from a repurposed photosystem, which now functions as a light-driven sensory mechanism.

The differences between prokaryotic and protist sensory capacities are not merely a matter of degree but represent access to a fundamentally different sensory world. Protists have developed more advanced and diverse sensory mechanisms, as well as more complex forms of motility, allowing them to inhabit a richer spatial and temporal domain. This complexity forces protists to solve more intricate detection-action coordination tasks.

This distinction becomes particularly evident when comparing eukaryotic and bacterial chemotaxis. Eukaryotic cells, being much larger than prokaryotes, can detect gradient differences by comparing asymmetric receptor activation across their membrane. This enables directed movement toward favorable targets or away from harmful areas. Unlike prokaryotes, eukaryotes organize locomotion through the synchronized activity of molecular motors in microtubules or actin filaments. Furthermore, eukaryotic cells process sensory information in ways that demonstrate rudimentary forms of 'learning' and 'memory', further distinguishing them from their prokaryotic counterparts (Fig. 4.2).

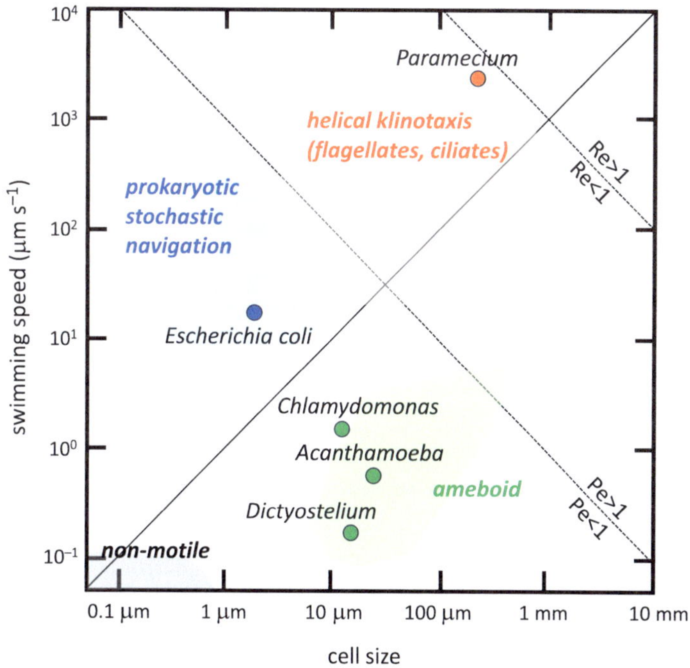

Fig. 4.2 Eukaryotes unlocked new biophysical spatio-temporal regimes unmatched by unicellular prokaryotes. There is a prokaryote-eukaryote divide with respect to behavior, as visualized in this phase space of organism size versus speed (log-log scale). Physical constraints partition this space, namely, the Reynolds number (Re, both unicellular prokaryotes and eukaryotes are dominated by viscosity), and the Péclet number (Pe, prokaryotes are diffusion limited and cannot reach Pe > 1). Adapted from Wan and Jékely (2021)

4.5 Agency in Associative and Multicellular Protists

It is widely accepted that the diverse evolutionary transitions from unicellular to multicellular have represented remarkable increases in complexity during the history of life. Integrated multicellular organisms are capable to harbor cellular differentiation and the formation of tissues and organs, and at the same time, they ensure a reliable cycle of development and reproduction of the whole multicellular body.

But this conversion of unicellular forms in multicellular bodies took many different forms, ending in more or less cohesive and integrated organizations. Albeit prokaryotic organisms have been capable to create relatively complex associations (see Chap. 3), eukaryotic organisms have developed a much more fascinating space between strict multicellularity and associative colonies. Strictly speaking, the transition towards true multicellular organisms starts with associations between eukaryotic cells, because they allow the emergence and evolution of much more complex

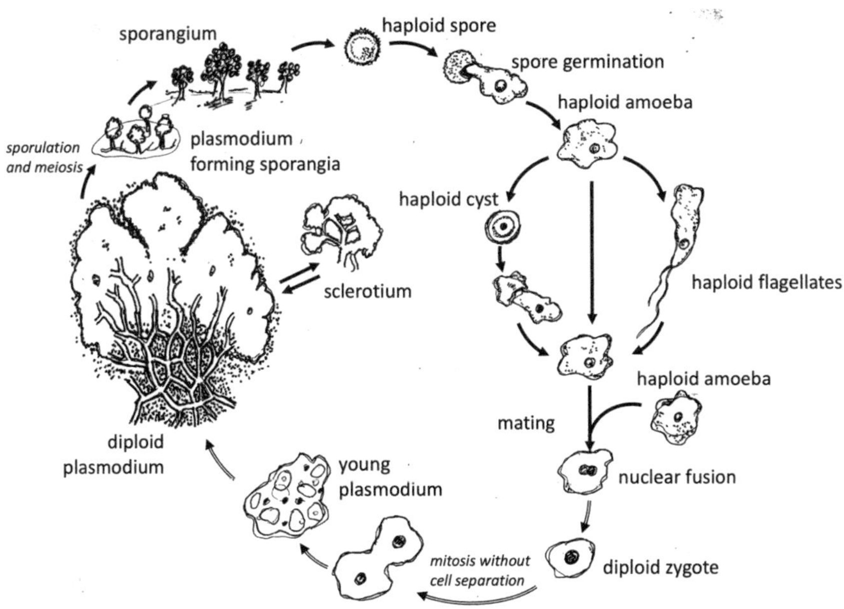

Fig. 4.3 The life cycle of *Physarum polycephalum*, a plasmodial slime mold, alternates between unicellular and multinucleated stages, adapting to environmental conditions. It begins with haploid spores that germinate into either flagellated or amoeboid cells. When conditions allow, compatible haploid cells fuse to form a diploid zygote, which grows into a multinucleated, syncytial (diploid) plasmodium through repeated nuclear division without cell division. The plasmodium actively seeks nutrients through chemotaxis, and can fragment or enter a dormant sclerotium state under unfavorable conditions. When resources become scarce, the plasmodium differentiates into sporangia, producing haploid spores through meiosis, thus completing the cycle. Illustration based on information given by Oettmeier et al. (2017) and Reid (2023)

and integrated collective systems. We will discuss in detail this transition in the next chapter.

We will now briefly examine three cases of eukaryotic multicellular organizations that exhibit notable forms of collective agency. However, these examples do not represent true or complex multicellular organisms. Instead, they are intermediate stages that have followed distinct evolutionary paths for various reasons. While many other protists display intriguing forms of collective behavior, we will focus on three specific examples: the coenocytic slime mold *Physarum polycephalum* (Fig. 4.3), the aggregative life cycle of the cellular slime mold (or social amoeba) *Dictyostelium discoideum* (Fig. 4.4) and the volvocine algae *Volvox* (Fig. 4.5).

Even if *P. polycephalum* is not strictly a multicellular system, its multinucleate structure provides valuable insights into how protists adapt their spatial exploration strategies as they grow in size. By contrast, *D. discoideum* undergoes a fascinating transformation during its life cycle, transitioning from a population of unicellular amoebae to a multicellular 'slug' and eventually to a fruiting body. This process illustrates a remarkable form of multicellular organization achieved through active

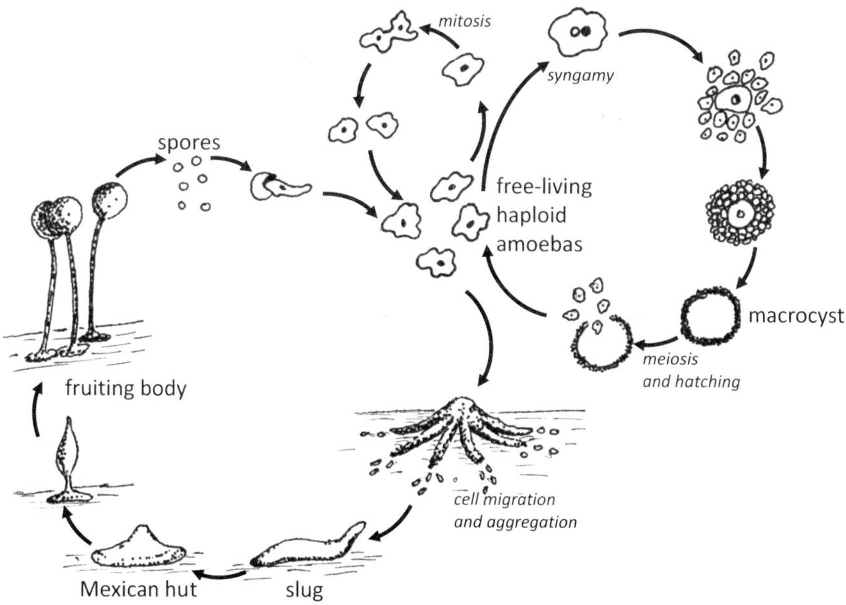

Fig. 4.4 The life cycle of *Dictyostelium discoideum*, a social amoeba, alternates between unicellular and multicellular stages, adapting to environmental conditions. In favorable conditions, free-living haploid amoeboid cells feed on bacteria and divide by binary fission. When food becomes scarce, thousands of amoebae aggregate in response to an extracellular cyclic AMP signal, forming a multicellular slug-like structure that moves toward light and heat. This slug eventually differentiates, through the stage of the so-called Mexican hut, into a fruiting body, containing haploid spores. These spores will disperse and remain dormant until conditions improve. Upon germination, spores release new amoebae, restarting the cycle. Under specific conditions, amoebae can also fuse to form a diploid macrocyst, which undergoes meiosis to generate genetic diversity. Illustration based on information given by Loomis (1975)

cell aggregation. Finally, the diverse species of *Volvox* offer a unique opportunity to compare the changes in agential motility between the presumed unicellular ancestors and their clonal multicellular descendants. In contrast, examples of true, more complex multicellularity—such as those seen in fungi, algae, and plants—will be explored in Chap. 5, while metazoan multicellularity will be discussed in Chap. 6.

4.5.1 Physarum polycephalum

The polynuclear coenocyte (or plasmodial) stage of *Physarum polycephalum* achieves a macroscopic size, enabling its large cytoplasmic volume to process sensory information and display a remarkable capacity for food searching. This behavior has been extensively studied and represents one of the most fascinating examples

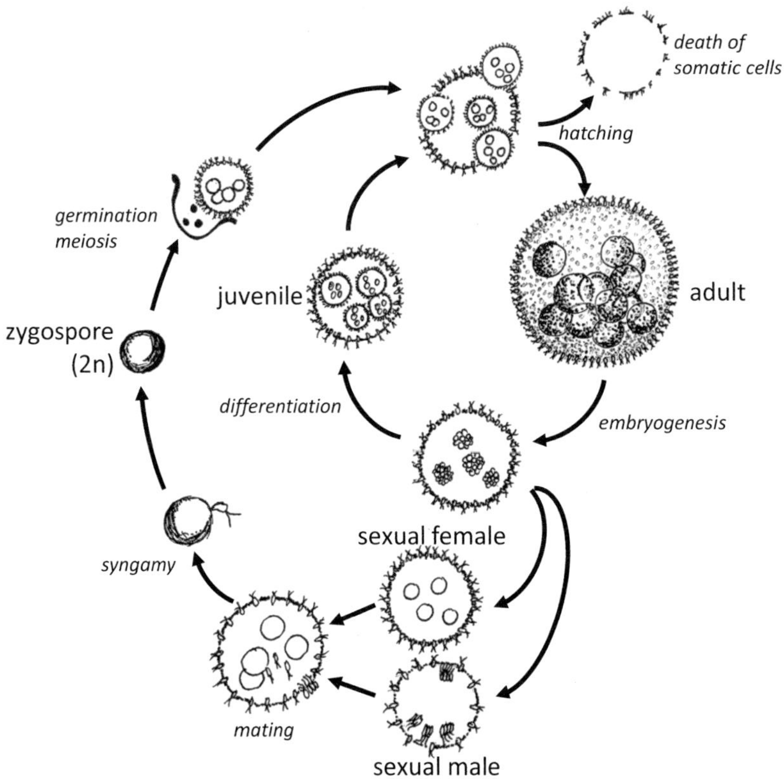

Fig. 4.5 The life cycle of *Volvox carteri*, a green alga, includes both asexual and sexual reproduction, adapting to environmental conditions. In favorable conditions, it reproduces asexually. Large reproductive cells called gonidia divide repeatedly to form juvenile spheroids within the parent colony. These juveniles grow flagella, invert to position them outward, and are eventually released as mature colonies, while the parent disintegrates. Under stress, such as nutrient depletion, *V. carteri* switches to sexual reproduction. Fertilization results in a diploid zygote that forms a thick-walled dormant zygospore, resistant to harsh conditions. When conditions improve, the zygospore undergoes meiosis, releasing haploid cells that form new colonies asexually. Illustration based on information in Umen (2020)

of behavior outside the realm of true multicellular organisms. Its learning and memory capacities are extraordinary, allowing this giant coenocyte to move in a slow yet flexible amoeboid manner, overcoming physical obstacles to locate optimal environmental conditions. Additionally, it can utilize information from past experiences to adjust its behavior.

At the plasmodial stage of its life cycle, *P. polycephalum* is capable of finding the shortest paths from a source vertex to all other vertices in a graph. This ability is hypothesized to rely on a network of interconnected vein-like tubes through which signaling molecules are transported. The organism leverages the large volume of its multinucleated cytoplasm to optimize its network's topological structure. These networks are modified by distinct oscillations that organize into spatiotemporal wave patterns, allowing the cell to efficiently evaluate and solve various situational challenges. All these oscillation cycles rely on a feedback system between the calcium signaling system, the contractile actomyosin network, and the energy metabolism.

P. polycephalum detects gradients of chemoattractants and repellents, responding to chemical and physical stimuli by altering patterns of electrical potential oscillations and contractions of its protoplasmic tubes. It optimizes its body configuration to enhance protoplasmic streaming, using thousands of biochemical oscillators that govern local clocks and propagate actin and calcium waves. Living in a world of gradients, the organism's responses are shaped by interactions with attractants and repellents. For example, while spreading on a substrate, it avoids regions with repellent concentrations above a certain threshold. This decentralized processing of sensory information translates into a spatially distributed configuration of protoplasmic tubes, resulting in directional movement and specific shape adaptations.

The adaptive behavior of *P. polycephalum* to environmental cues, such as light, chemical signals, or physical barriers, involves a complex interplay between movement and growth. These two processes are distinct yet interconnected. When encountering a food source, it extends its pseudopodia toward the food, demonstrating both movement (adaptive behavior) and growth (physical expansion). This extension involves adding new cellular material, increasing the organism's overall size. Movement, driven by cytoplasmic streaming, allows *Physarum* to explore and navigate its environment. While this movement does not necessarily involve growth, the organism can consolidate resources by retracting pseudopodia from unproductive areas and redirecting them toward more favorable regions.

This dynamic reallocation exemplifies adaptive behavior, leading to growth in the direction of the food source. *Physarum* forms and reconfigures networks to optimize nutrient uptake. When a food source is detected, it can grow new veins toward the food while retracting from less favorable areas. This process combines adaptive behavior (network reconfiguration) and growth (vein extension). Adaptive behaviors may involve temporary changes, such as reversible cytoplasmic streaming, or permanent alterations, such as growth toward a nutrient source. Ultimately, the organism's adaptive capacity lies not only in growth but also in its ability to change shape, move, and reallocate resources dynamically.

4.5.2 Dictyostelium discoideum

Dictyostelium discoideum exhibits a life cycle that alternates between a free-living single-celled stage and a multicellular stage before spore formation. The multicellular stage is fundamental to the life cycle, as reproduction can occur in two ways: either as unicellular organisms through mitosis (asexual reproduction) under favorable conditions, or under adverse conditions through aggregation or fusion of unicellular individuals. This aggregation results in a macroscopic, multicellular motile entity that culminates in the formation of a fruiting body. The fruiting body dies after releasing spores, which disperse and, upon finding suitable conditions, initiate a new life cycle.

One intriguing aspect of *D. discoideum* is that its agency manifests in two distinct forms: one during the individual stage and the other during the associative stage. In the latter, collective motility plays a crucial role, making this form of protist collective agency particularly noteworthy. When food (typically bacteria) is abundant, dictyostelids behave as individual amoebae, feeding and dividing as usual. However, when food is depleted, they aggregate to form a multicellular assembly capable of migration. Under appropriate conditions, this assembly matures into a fruiting body, consisting of a stalk and spores. The spores, which are dormant cells with resistant walls, eventually germinate into new amoebae once food becomes available.

Under normal conditions, unicellular amoeboid eukaryotes move slowly and within limited ranges. However, in hostile environments, they achieve long-distance migration by aggregating into a transient multicellular entity. During this process, approximately 20% of the cells differentiate and die to form the stalk, while the remaining cells differentiate into spores. These spores are released into the air and dispersed by the wind to establish new populations. At this stage, the collective entity no longer moves itself but relies on external forces (e.g., wind) for spore dispersal.

Alternatively, one could describe the unicellular amoeboid cells as displaying diverse forms of action. When faced with unfavorable individual conditions, they shift to cooperative behavior, resulting in a new functional action: migration to a better environment. However, it becomes challenging to attribute this behavior to individual unicellular entities, as this collective behavior is a property of the transitory multicellular system and persists beyond the lifespan of many individual amoebae (many of which die during the process). Experimental data and computational simulations indicate that the extracellular matrix surrounding the mobile collective of cells constitutes a flexible dissipative self-assembled structure whose flexibility plays a critical role in the movement of the 'slug'.

The plastic and complex extracellular matrix of *D. discoideum* plays a crucial role not only in regulating cell motility but also in supporting multicellularity, cell differentiation, and morphogenesis. It provides both structural support and protective layers for the resulting differentiated cell types. Throughout migration and development, the extracellular matrix undergoes dynamic remodeling, with its composition and properties influenced by the cell's actomyosin-based contractility.

A useful comparison can be made between *D. discoideum* and the multicellular form of the prokaryote *Myxococcus xanthus*. In *M. xanthus*, coordinated multicellular motility is facilitated by an ECM-like structure formed from secreted materials. These secretions create physical pathways or trails that guide the movement of other cells, which follow the tracks established by earlier cells. This form of coordination relies on stigmergic regulation—where individual actions modify the environment in ways that influence subsequent behavior. In contrast, *D. discoideum* coordinates collective behavior through active degradation and remodeling of its extracellular matrix by proteases, which in turn modulates cell motility.

The motility of the *D. discoideum* slug is considerably more complex than that of *M. xanthus*. Cell size and stiffness can hinder faster-moving cells from bypassing slower ones, introducing physical constraints on movement. The overall patterns of migration differ significantly depending on the relative adhesiveness of prestalk and prespore cells. Variations in both the strength and type of cell adhesion between different cell types play a major role, and the collective movement of the slug is fundamentally three-dimensional.

As Niklas and Newman have noted, the individual amoebae within the *D. discoideum* slug behave like a cohesive liquid. Actomyosin-driven cellular mobility allows a kind of collective fluid-like behavior embedded in a cellulose-based extracellular matrix that is both "compliant and permissive" to cell rearrangement, enabling a high degree of coordinated multicellular motion.

In summary, slime molds like dictyostelids defy any simple classification in terms of individual agents. A significant part of their maintenance and reproductive strategies relies on the temporary associative aggregate entity they form during their life cycle.

4.5.3 Multicellular Volvocine Algae

Volvox species are multicellular eukaryotic green algae that inhabit various freshwater environments. Each organism may consist of up to 50,000 cells, organized into two distinct cell types: a large number of flagellated somatic cells and a smaller number of germ cells. These cells are embedded in the surface of a hollow sphere supported by an extracellular matrix. The somatic cells specialize in movement and energy production for the colony, while the germ cells, rich in resources, are dedicated to reproduction. This division of labor enhances both the metabolic efficiency and reproductive potential of the multicellular entity as a whole.

Adult somatic cells in *Volvox* form a single outer layer, with their flagella oriented outward, enabling efficient collective motility. To achieve rapid swimming, individual flagella move in a highly coordinated manner, allowing the entire multicellular colony to propel itself effectively. In addition, somatic cells possess eyespots that guide *Volvox* toward light, facilitating phototactic movement through the synchronized action of thousands of flagellated cells.

In response to light stimuli, both the flagellar waveform and beat frequency adjust, with the effective stroke direction reversing as needed. A gradient in photoresponse extends from the anterior to the posterior pole of the spheroid. The anterior hemisphere, which has larger and more numerous eyespots, is highly responsive to light, while the posterior hemisphere, with minimal or no eyespots, remains largely "blind." When illuminated, anterior cells reverse their flagellar beating direction, causing the colony to turn toward the light source. Gravitational forces further support positive phototaxis, enhancing *Volvox*'s ability to navigate its environment efficiently.

Detailed studies of phototactic behavior in *V. rousseletii* reveal a photophobic response to sudden changes in light intensity. Initially, the multicellular entity decelerates for approximately 2 s due to flagellar reversals in anterior cells with large, sensitive eyespots, creating a temporary sink. These cells then resume forward beating, accelerating the spheroid and directing it phototactically toward the light source. The anterior cells are highly photosensitive, while the posterior cells are unresponsive, resulting in a gradient of photosensitivity. As the spheroid rotates counterclockwise, the uneven flagellar forces between illuminated and shaded sides steer it toward the light. This collective movement arises from individual cell responses to environmental light and gravitational cues.

Volvox likely evolved from a unicellular, free-swimming ancestor similar to extant *Chlamydomonas* species, which possess an eyespot and flagella to navigate light gradients and optimize radiant energy absorption. Comparison of the genomes of *C. reinhardtii* and *V. carteri* indicates that both contain the same number of protein-coding genes. This fact offers an extraordinary example of how multicellular complexity is not the result of a net increase in genome size but of the amplification and diversification of specific families of genes (for example, those related to the extracellular matrix or the regulation of the cell cycle).

The transition from a unicellular *Chlamydomonas*-like ancestor to the multicellular *Volvox* involved several key innovations, including the coordinated movement of all cells' flagella, the development of an extracellular matrix to structurally cement the multicellular entity, a strict genetic control of cell differentiation into somatic and germ cells, and the evolution of light-stimulated, transient ciliary-type asymmetrical beating to move in reverse directions, replacing flagellar-type symmetrical beating. This set of correlated innovations, among others, enabled a great diversity of *Volvox* species to successfully thrive as multicellular organisms in aquatic habitats.

4.6 Protist Agency Is Amazing But Has Its Limitations

As discussed in this chapter, the rapid diversification of protists and their remarkable behavioral innovations may represent the earliest clear manifestation of agency complexification in the history of life. Protists have capitalized on the internal mechanisms established during eukaryogenesis, allowing them to benefit from larger body sizes compared to prokaryotes. Despite the energetic and informational costs associated with their increased size, many protists have significantly enhanced their interactive capabilities for outward action. These innovations—such as advanced movement, sensory organ development, and the ability to process and integrate environmental information—have enabled more complex behaviors, facilitating their colonization of an extraordinary range of ecological niches.

However, assessing the significance of this agency complexification is challenging, as there is no unique evolutionary trajectories toward more complex forms of agency. For example, while *Paramecium* relies on rapid motility, dinoflagellates have developed sophisticated sensory systems, and amoebozoans, with their slower yet highly adaptable movement, exhibit learning capabilities as demonstrated in *Physarum*. Some protists even shift interaction modes during their life cycles. Yet, if we focus on these cases, it is clear that they represent remarkable advancements in agency, as compared to prokaryotic cases.

This highlights that the complexification of agency is not inherently advantageous, prompting the question: why did some protists evolve more complex forms of agency than others? One explanation lies in their effective use of eukaryotic internal machinery to explore new ecological niches and develop novel modes of interaction. Agency complexification refers to the degree and sophistication of the interactive strategies an organism employs to survive and reproduce.

Depending on their ecological context, some protists require greater energy and information-processing capabilities to carry out relatively complex behaviors, while others do not. For instance, planktonic diatoms—highly successful ecologically—lack intrinsic motility and rely on ocean surface currents to remain suspended in sunlit waters. Despite this limitation, they can regulate their buoyancy using ion pumps. At the opposite end of the spectrum are bacterivorous ciliates like *Spirostomum ambiguum*, which have evolved remarkably complex sensorimotor systems to actively engage with their environment.

Thus, it is not uncommon to find protists in which internal biological complexity has not translated into more complex forms of agency, simply because they have evolved into relatively passive life forms. For example, in planktonic diatoms—organisms primarily transported by ocean currents—the actomyosin machinery is used more for constructing their intricate silica-based cell walls than for supporting active behavioral functions.

Therefore, to understand the factors driving the complexification of agency, it is not sufficient to examine the evolution of an organism's internal or structural complexity alone. Instead, we must focus on their specifically interactive functions and investigate the conditions under which selective pressures act on the mechanisms that support these functions—such as sensory and motor systems.

In abstract terms, agency is the ability of an autonomous agent to modify its environment in a functional way. To do so, the agent must evaluate environmental variables and modulate its actions accordingly, implementing what is known as an "action-detection loop". In a biological framework, the potential for complexifying these loops depends on two critical factors: (1) whether the action involves secretion or motility, and (2) the size of the agent, since larger agents require more complex internal machinery for coordinated behavior. These factors are crucial for understanding the evolution of agency in protists. We will now analyze how motility and size may have driven this complexification.

4.6.1 Motility as a Driver of Complexification

The mode of interaction—whether through secretion or movement—influences the ways in which agency can evolve. Motility-based agency opens up different possibilities for complexification compared to secretion-based agency. Mobile organisms can actively adjust their direction and speed, enabling the spatial repositioning of sensors. This mobility allows for the integration of diverse environmental inputs, such as light, sound, and chemical gradients, thereby broadening the range and responsiveness of interactions. In contrast, secretion-based agents influence their surroundings primarily through the modulation of secretory outputs. While this can support sophisticated forms of environmental engagement, it generally offers a more constrained set of sensory-effector correlations.

Motility may facilitate the evolution of advanced sensory systems, such as eyes, which enable dynamic perception of objects and spatial relationships. This enhances the richness of sensory-motor coordination. Although secretion-based systems—like those found in plants (discussed in Chap. 5)—can also reach high levels of evolutionary complexity, they tend to operate with less flexibility and immediate responsiveness than mobile systems.

The diversification of protists suggests that the emergence of complex sensors, which are crucial for more elaborate behaviors, is often linked to motility. Flexible and responsive movement appears to support the development of new functional structures and behaviors, offering a pathway for the increasing complexity of agency.

4.6.2 Size as a Driver of Complexification

A second major factor in the complexification of agency is the size of the agent. As seen in this chapter, the increased size of protists has enabled and constrained innovations in agency. While motility provides a foundation for complexification, larger size allows for more integrated internal machinery, supporting fine-tuned actions, precise sensory systems, and advanced information processing.

However, larger size also presents challenges for maintaining fast and flexible motility. Analyzing the evolution of agency in protists requires considering the interplay between size and motility, acknowledging the trade-offs and complementarities. Since multicellularity is the primary solution for achieving larger size (as the limitations of the coenocytic form show), studying multicellular protists is crucial for understanding the extent of agency complexification.

4.6.3 Assaying Multicellular Agency in Protists

Multicellular forms introduce novel capacities that transcend the limitations of single cells. By enabling organisms to scale in size independently of cell size, multicellularity frees them from the constraints of unicellular existence, facilitating new interactions with physical and biological environments. However, multicellular protists face significant limitations in their ability to diversify and complexify. Aggregative multicellularity, for instance, struggles to avoid intercellular conflicts. In coenocytic forms, internal differentiation is difficult to achieve. Even in large multicellular protists, such as *V. carteri* (up to 50,000 cells) and *D. discoideum* (aggregating up to 100,000 cells), coordination mechanisms limit behavioral complexity. For example, *D. discoideum* uses extracellular cyclic AMP for triggering cell aggregation, while multicellular volvocine algae, despite being clonal, lack advanced intercellular signaling. In contrast, the rosette colonies of the choanoflagellate *Salpingoeca rosetta* (with a very reduced number of cells) exhibit signs of an evolutionary toolkit for behavioral complexification. These colonies feature clonal cells with greater differentiation and coordination, regulated by electrical signals and calcium ion waves associated to a form of voltage-gated calcium channel.

The colonial stage of *S. rosetta* shows how the emergence of complex agency is shaped by a combination of factors. These include its ability to coordinate a form of multicellular organization that, although smaller in scale than that of *D. discoideum* or multicellular volvocine algae, may be more differentiated and integrated. This organization also enables faster and more flexible movement, made possible by the synchronization of individual cells through shared electrical signaling pathways. Together, these features represent a significant step along the evolutionary path toward more complex forms of agency.

4.6.4 Final Remarks

The emergence of eukaryotic cells allowed evolution to explore new ecological and organizational possibilities. While increased complexity is just one of many evolutionary pathways, protists demonstrate diverse strategies for agency complexification. Through the evolution of different modes of agency, protists have profoundly influenced their evolutionary trajectories. Even fungi and plants, which represent advanced eukaryotic complexity, trace their origins to mobile protists. The unfolding of agency requires a deep process of co-evolution with developmental and physiological mechanisms, shaping the history of eukaryotic life. This chapter has explored the diverse paths and limitations of protist agency complexification. In the following chapters, we will examine how these constraints have shaped the evolution of more advanced forms of complexity.

Bibliography

Arnellos A, Moreno A (2015) Multicellular agency: an organizational view. Biol Philos 30:333–357. https://doi.org/10.1007/s10539-015-9484-0
Barandiaran X, Moreno A (2009) Adaptivity: from metabolism to behavior. Adapt Behav 16:325–344. https://doi.org/10.1177/1059712308093868
Bonner JT (2006) Why size matters: from bacteria to blue whales. Princeton University Press, Princeton, NJ. https://www.jstor.org/stable/j.ctt7s5hv
Boussard A, Fessel A, Oettmeier C et al (2021) Adaptive behaviour and learning in slime moulds: the role of oscillations. Phil Trans R Soc B 376:20190757. https://doi.org/10.1098/rstb.2019.0757
Loomis W (1975) *Dictyostelium discoideum*: a developmental system. Elsevier, Amsterdam
Miyata M, Robinson RC, Uyeda TQ et al (2020) Tree of motility–a proposed history of motility systems in the tree of life. Genes Cells 25:6–21. https://doi.org/10.1111/gtc.12737
Niklas KJ, Newman SA (2020) The many roads to and from multicellularity. J Exp Bot 71:3247–3253. https://doi.org/10.1093/jxb/erz547
Nilsson D-E, Marshall J (2020) Lens eyes in protists. Curr Biol 30:R458–R459. https://doi.org/10.1016/j.cub.2020.01.077
Oettmeier C, Brix K, Döbereiner HG (2017) *Physarum polycephalum*—a new take on a classic model system. J Phys D Appl Phys 50(41):413001. https://doi.org/10.1088/1361-6463/aa8699
Reid CR (2023) Thoughts from the forest floor: a review of cognition in the slime mould *Physarum polycephalum*. Anim Cogn 26:1783–1797. https://doi.org/10.1007/s10071-023-01782-1
Richards TA, Eme L, Archibald JM et al (2024) Reconstructing the last common ancestor of all eukaryotes. PLoS Biol 22:e3002917. https://doi.org/10.1371/journal.pbio.3002917
Ueki N, Matsunaga S, Inouye I et al (2010) How 5000 independent rowers coordinate their strokes in order to row into the sunlight: phototaxis in the multicellular green alga *Volvox*. BMC Biol 8:103. https://doi.org/10.1186/1741-7007-8-103
Umen JG (2020) *Volvox* and volvocine green algae. EvoDevo 11:13. https://doi.org/10.1186/s13227-020-00158-7
Vosseberg J, van Hooff JJE, Köstlbacher S et al (2024) The emerging view on the origin and early evolution of eukaryotic cells. Nature 633:295–305. https://doi.org/10.1038/s41586-024-07677-6
Wan KY (2024) Biophysics of protist behaviour. Curr Biol 34:R981–R986. https://doi.org/10.1016/j.cub.2024.07.002
Wan KY, Jékely G (2021) Origins of eukaryotic excitability. Phil Trans R Soc B 376:20190758. https://doi.org/10.1098/rstb.2019.0758

Chapter 5
Multicellular Agents: The Challenge of Combining Body Organization and Behavior

Abstract This chapter explores the evolution of complex multicellularity and its implications for organismal agency, with a focus on sessile multicellular lineages. Although multicellularity has evolved independently several times, only a few lineages have reached high levels of organizational complexity. The chapter analyzes the structural and functional differences between sessile and motile integrated multicellular organisms, highlighting the significance of cell walls in non-motile multicellular life and the evolution of specialized tissues in plants and fungi. It examines the constraints faced by sessile organisms and their adaptive strategies, particularly their reliance on secretory interactions and plastic growth responses to environmental stimuli. Ultimately, the chapter demonstrates that an increase in structural complexity does not necessarily correspond to a parallel increase in the complexity of agency.

Keywords Complex (true) multicellularity · Sessile organisms · Plant evolution · Fungal evolution

5.1 Introduction

At the end of the previous chapter, we saw that eukaryotic cells readily form a diverse array of multicellular systems. In fact, the production of polysaccharides and structural glycoproteins that self-assemble to form extracellular matrices facilitates the multicellular associations of eukaryotic cells. These extracellular matrices provide structural support and promote the stability of multicellular arrangements. As a result, the eukaryotic tree of life exhibits a remarkable diversity of colonial, aggregative, and clonal multicellular entities.

From an evolutionary perspective, the emergence of eukaryotic multicellularity seems almost inevitable and simple multicellular forms are present in the fossil

Á. Moreno, J. Peretó, *An Evolutionary Story of Agency*, Foundations for Interdisciplinarity in the Life Sciences: Concise Monographs,
https://doi.org/10.1007/978-3-032-05644-3_5

record as early as 1.6 billion years ago. Multicellularity likely became an obligatory step in the life cycle or developmental program of certain lineages, where a single cell reproduces clonally and gradually gives rise to differentiated cells. This transition is significant because natural selection begins to act on the integrated whole rather than on individual cells, making the entire system a unit of selection.

Development enables the formation of a functionally integrated organism from initially independent cells, allowing the entity to interact with its environment as a single unit. From the outset of this process, cells synchronize their activities, forming a coordinated and functionally cohesive system. However, for a fully developed multicellular organism to emerge, several evolutionary innovations are required—such as cell-cell adhesion, intercellular communication, long-range transport mechanisms, and programmed cell death. As cells differentiate, they must also organize into specialized structures (e.g., tissues), which help maintain multicellularity by preventing a return to a unicellular state. Ultimately, this process results in the emergence of species-specific morphologies and an irreversible differentiation of cell types.

At the same time, the developmental program enables initially undifferentiated cells to acquire new functions and form increasingly complex multicellular structures. However, as the number and functional diversity of cells grow, maintaining integration becomes more challenging. This necessitates additional organizational mechanisms, which is why, as we shall see, different forms of multicellularity have evolved varying levels of complexity throughout the history of life. In summary, the emergence of multicellular organisms required numerous evolutionary innovations, all of which depended on the development of an integrated intercellular organization.

5.2 Complex Multicellularity

Despite the diversity of multicellular forms, eukaryotic cells have achieved only a few types of multicellular organization capable of following a true evolutionary trajectory toward complexity. "True" or "complex" multicellularity is rare across the tree of life because lineages must overcome numerous genetic and developmental obstacles to reach this stage. Many biologists consider complex multicellularity to be present in only six major groups: brown and red algae, basidiomycete and ascomycete fungi, land plants, and animals (Fig. 5.1). This classification has led researchers to identify a set of distinguishing characteristics shared exclusively by organisms within these groups, which are currently recognized as true multicellular organisms.

According to Andrew Knoll, the key criterion for determining whether a multicellular system qualifies as a true multicellular organism is whether all of its cells are in direct contact with the environment. In complex multicellularity, some cells interact only with other cells, meaning they cannot acquire oxygen and nutrients solely through diffusion. This necessitates the evolution of specialized transport

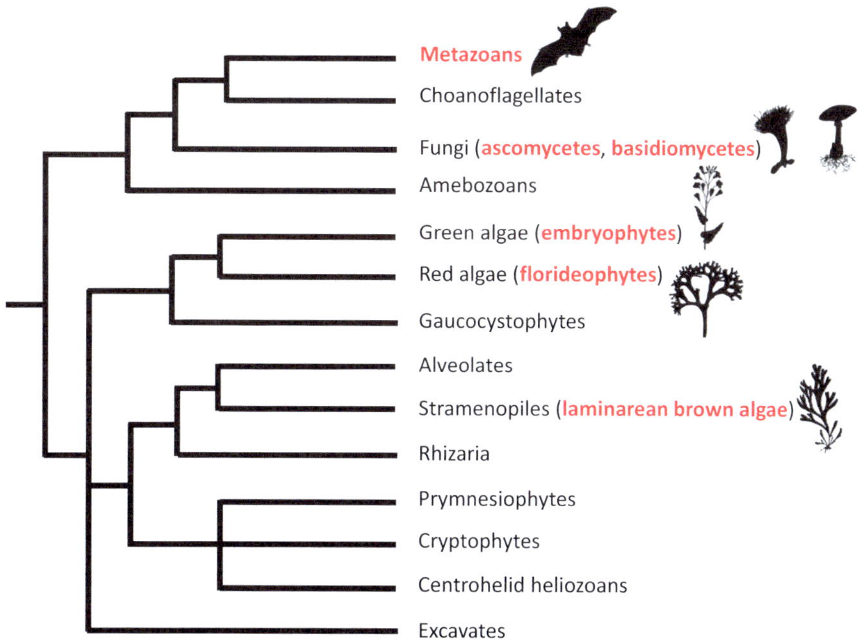

Fig. 5.1 Eukaryotic phylogeny based on Fig. 1 by Knoll (2011) showing the position of the six true multicellular organisms: metazoans, ascomycetes, basidiomycetes, embryophytes (land plants), florideophytes (red algae), and brown algae. The silhouettes correspond to (from top to bottom, and left to right) *Myotis nattereri*, *Zodiomyces vorticellarius* (ascomycota), *Amanita phalloides* (basidiomycota), *Brassica rapa*, *Chondrus crispus*, and *Fucus serratus* reproduced under a Creative Commons License from Phylopic (https://www.phylopic.org/)

mechanisms to overcome such limitations. In line with this, Leonardo Bich and colleagues have proposed that a defining principle of integrated, organismal multicellularity is the functional organization of intercellular space. The ability to organize in space enables cell differentiation, an increase in body size, and even specialized functions such as immunity. The extracellular matrix plays a crucial role in this process, acting as an evolutionarily ancient, non-cellular control system that helps regulate the functional organization of multicellular space, influencing cell fate and behavior. Additionally, complex multicellular systems exert control over internal movement and communication. Similarly, other authors emphasize that a complex multicellular, macroscopic body, with multiple cell types, is the outcome of a developmental program that orchestrates cell division and specialization.

In sum, an integrated multicellular organism requires coordinated intercellular control to maintain proper function. This includes regulating the number and spatial distribution of different cell types, managing cell death (apoptosis), and the removal of dead cells, ensuring effective intercellular transport of substances, and defending against pathogens. In other words, a form of housekeeping control is essential to maintain both the material and operational aspects of global homeostasis.

Although the six major lineages of multicellularity are all considered true or complex, they have evolved vastly different degrees of organizational integration. Their evolutionary histories reveal variations in key aspects such as the extent of cellular differentiation, the capacity for long-range intercellular control, and the sophistication of their developmental programs.

This chapter will outline the main types of true multicellular organisms that have emerged throughout the history of life, examining their unique characteristics and evolutionary trajectories. We will explore why each lineage has evolved differently and how they have developed varying levels of structural and functional complexity. While we distinguish between true and non-true multicellularity, it is important to recognize that true multicellularity itself manifests in different degrees of complexity and integration.

We will focus on how different organizational structures either enable or constrain evolutionary innovations and increasing levels of biological integration. We will also examine how different forms of multicellular organization shape their interactions with the environment. However, the specific case of metazoans will be addressed in the following chapter. Through this discussion, we aim to clarify the rationale behind our classification of multicellular organisms and the broader evolutionary patterns that have shaped their diversity.

5.3 Two Groups of Multicellular Organisms: Sessile and Motile

The six forms of eukaryotic complex multicellularity have emerged from distinct organizational and ecological contexts. To explain these diverse evolutionary paths, we will consider two key aspects: the constitutive characteristics of their respective evolutionary ancestors, and the ecological circumstances in which they evolved.

5.3.1 Constitutive and Organizational Factors

From an organizational perspective, multicellular algae, plants, and fungi share a common evolutionary constraint that has shaped their multicellular development: they have all lost locomotion during their development and maturity, restricting movement to their germinal cells. In these lineages, the evolution of complex multicellularity required the emergence of new cell types that, despite lacking motility, could undergo functional differentiation to form tissues.

A key adaptation in these sessile multicellular organisms is the development of specialized cell walls that provide structural integrity and facilitate intercellular interactions. In multicellular algae and plants, the cellulose-based cell wall offers mechanical support for growth while also serving as a protective barrier against pathogens. At the same time, it must remain flexible enough to remodel in response to developmental and environmental cues. In multicellular fungi, the chitin-based cell wall regulates cellular interactions with the environment, providing rigidity and protection against osmotic stress. Notably, the presence of a cell wall has a significant biophysical consequence: it enables cells to withstand internal turgor pressures of up to 10 atmospheres, which can be harnessed for growth and movement, despite limiting motility.

In contrast, metazoans are inherently motile organisms composed of contractile, wall-less cells that migrate during development and throughout adult life. The absence of a rigid cell wall allows for greater cellular mobility and dynamic tissue formation, both of which are essential for their motile way of life.

One might ask why non-locomotive, cell wall-based multicellular lineages (five cases) are more numerous than the motile, wall-less lineage of metazoans. Moreover, the majority of these non-motile lineages exhibit relatively low complexity compared to animals. For instance, the number of distinct cell types in algae and fungi is typically fewer than 12, while in plants, which are the most complex non-motile multicellular organisms, the number reaches around 30. In contrast, animals can have hundreds of different cell types. This disparity suggests that the cell wall-based pathway may provide a simpler evolutionary route to multicellularity than the more complex, motile, wall-less pathway. If this is true, the loss of motility in somatic cells might be the most straightforward way to achieve a stable and integrated multicellular organization.

5.3.2 Ecological and Evolutionary Context

Another crucial factor shaping these differences is the ecological niche within which each form of multicellularity evolved. Multicellular algae and plants evolved as autotrophic, sessile organisms, relying on photosynthesis and remaining fixed in one location. Fungi are heterotrophic organisms that have adapted to a sessile lifestyle, secreting enzymes and decomposing external organic matter before absorbing it into their cells. Metazoans, by contrast, evolved as heterotrophic, motile organisms that actively search for and internally digest organic material. These fundamental ecological distinctions have significantly influenced how each lineage developed its multicellular organization and complexity. As we shall see, these differences have resulted in major variations in organismal complexity across evolutionary history.

From their earliest evolutionary origins, adaptations to either a sessile or a motile lifestyle have shaped the structural organization of multicellular organisms. As these organisms moved toward increasingly integrated multicellularity, their organizational structures became more specialized, and evolutionary pressures reinforced these constraints. Each type of multicellular system developed under distinct conditions that determined its evolutionary trajectory, as natural selection acted upon the possibilities offered by its specific organizational framework. As a result, the fundamental structure of an organism's multicellular organization profoundly influences its interactive evolutionary potential.

This raises an important question: to what extent do the degree of structural (constitutive) complexity and the degree of interactive complexity correspond to one another? Are there fundamental asymmetries between these two dimensions? To explore this, we will first examine the different levels of constitutive complexity in the five forms of sessile multicellular organisms and the factors that shaped their evolution. As mentioned above, the case of metazoans, which represents a distinct evolutionary trajectory, will be addressed in the following chapter.

5.4 Comparative Analysis of the Constitutive Complexity of Sessile Multicellular Organisms

Fungi, algae, and plants have evolved distinct levels of multicellular complexity due to different evolutionary and environmental factors. Despite these differences, all three groups have been highly successful, colonizing a wide variety of ecological niches, including terrestrial environments. Below, we will examine the specific characteristics of fungal and algal/plant multicellular organization. For simplicity, we will treat Basidiomycota and Ascomycota as a single group.

5.4.1 Fungal Multicellular Organization

Unlike most multicellular organisms, most of the basidiomycota and ascomycota fungi are best understood as clonal multicellular organisms, with clonality occurring at the level of individual nuclei rather than entire cells. This distinction arises from fungi's unique filamentous growth and the evolutionary mechanisms that led to their multicellularity.

How did unicellular ancestors transition to multicellular fungi? Fungi originated from a group of aquatic unicellular members of the Opisthokonta supergroup (constituted by metazoa, fungi and their closest unicellular relatives). Some of these ancestors evolved into motile parasites, while others adopted a saprotrophic lifestyle. In the latter group, the loss of flagella was a significant evolutionary shift. This loss is linked to the presence of a cell wall throughout all stages of their life cycle, except in their germinal stage as spores.

A key factor in this transition was the evolution of the chitin-glucan cell wall, which played a crucial role in controlling nutrient absorption. This wall also provided protection from osmotic stress, structural rigidity, and defense against fungal pathogens. However, the most significant innovation in fungal multicellularity was the development of hyphal growth.

A hypha is a filamentous structure composed of multiple cells enclosed within a tubular fungal cell wall. In multicellular fungi, hyphae are divided by transverse septa, which contain large pores that allow ribosomes, mitochondria, and even nuclei to move between cells. Hyphal growth played a crucial role in fungal evolution, enabling these sessile heterotrophic organisms to extend hyphae in response to specific chemical cues. The transition to multicellularity occurred as fungi adapted to terrestrial environments, where nutrients were unevenly distributed. This shift likely required a foraging strategy capable of efficiently accessing distant resources. Given these conditions and the structural constraint of a rigid cell wall, the development of an elongating, filamentous thallus emerged as the most effective—or possibly the only—solution, driving the evolution of hyphae.

Unlike coenocytic single-celled organisms, where the cytoplasm is continuous, multicellular fungi use septa to compartmentalize hyphae. These septa have perforations that enable cellular organelles and even nuclei to migrate between compartments. Hyphal growth occurs through the extension of cell walls and internal components, with new septa forming as the hyphae elongate. This growth pattern allows for efficient branching, where new hyphae extend from existing tips, enabling fungi to rapidly explore their environment.

Hyphal growth is tightly regulated by a network of extracellular signals that direct polarized growth, initiate transcriptional programs, and prepare cells for membrane fusion and formation of multinucleated, genetically heterogeneous cell. Fungal multicellularity is based on networks of hyphae, which can form specialized tissues, namely, the plectenchyma—loosely entangled hyphae forming flexible networks—and the sclerotial tissue—dense, compact masses of hyphae providing structural support. Hyphae also exhibit functional specialization. For instance, in parasitic fungi, hyphae are adapted for penetrating host tissues and absorbing nutrients. On the other hand, many fungi establish symbiotic relationships with plants (mycorrhizae), where fungal hyphae interface with plant roots.

Fungal multicellularity evolved along a distinct trajectory compared to other multicellular lineages. Instead of following the typical sequence of adhesion, communication, and cell differentiation, fungi developed multicellularity through key innovations in hyphal elongation, compartmentalization, communication, and differentiation. Unlike most lineages, fungi did not initially rely on cell adhesion as the foundation for multicellular organization. Although adhesion became important later—particularly in the development of complex structures like fruiting bodies—early fungal hyphae likely arose through mechanisms other than simple adhesion. This unique evolutionary pathway allowed fungi to exploit a wide range of ecological niches, profoundly influencing the structure and function of modern ecosystems.

5.4.2 Plant and Algae Multicellular Organization

Land plants, along with brown and red algae, are three distinct groups of photosynthetic multicellular organisms. Green and red algae, as well as land plants, originated from an early endosymbiotic event between a primitive photosynthetic cyanobacterium and an ancestral eukaryotic phagotroph, whereas brown algae evolved through a secondary endosymbiosis involving a red alga and a heterotrophic protist ancestor.

The closest relatives of plants are streptophyte green algae, with Zygnematophyceae identified as their sister group. Compared to multicellular algae, plants have evolved significantly greater complexity. Algae lack key plant structures such as vascular tissues, true shoots, stems, leaves, and roots. In contrast, plants exhibit a highly diversified tissue system, including xylem and phloem, which form the vascular network responsible for transporting water, nutrients, and organic compounds throughout the organism. They also possess specialized organs such as roots (water and mineral absorption), stems (support and transport), leaves (photosynthesis), and flowers (reproduction). Given these significant differences, our discussion will hereafter focus on plants.

Although plants originated in aquatic environments, they evolved to colonize land, leading to the emergence of major plant groups and distinct cell types (e.g., xylem and phloem in vascular plants). This transition had a profound impact on the biosphere, paving the way for terrestrial animal life. The move from an aquatic environment to land was driven by key innovations, including symbiotic associations with mycorrhizal fungi, the development of stress-response genetic toolkits, and modifications in the cell wall.

Recent studies have also identified fundamental genomic innovations underlying the origin of land plants, particularly those enabling embryogenesis. Notably, after embryogenesis, the mature embryo does not reflect the adult plant's final structure. Instead, undifferentiated stem cells within meristems continue to drive development, giving rise to various plant cell types and tissues. Unlike animals, plants lack a defined germline, and their multicellularity is extremely plastic characterized by continuous development, together with flexible and reversible cellular differentiation.

In the same way as for fungi, the multicellularity of algae and plants also depends on specialized cell walls. Brown algal cell walls are primarily composed of polyanionic polysaccharides such as alginates and fucose-containing sulfated polysaccharides, with cellulose playing a secondary role. Red algae possess a double-layered cell wall: an outer layer rich in polysaccharides and an inner layer composed mostly of cellulose.

Plant cell walls, however, are structurally more complex, containing cellulose, hemicellulose, and pectin. The cellulose fibrils in plant cell walls provided the earliest land plants with protection against environmental stressors. Beyond structural support, plant cell walls serve as physical scaffolds that hold cells together, enabling tissue differentiation. Over evolutionary time, plant cell walls have acquired additional functions, including defining mechanical properties, maintaining cell morphology, facilitating intercellular communication, regulating cell differentiation, and influencing physiological processes. Notably, plant cell walls contain microscopic channels (plasmodesmata) that enable transport and communication between cells. As mentioned before, the capacity of cell wall to sustain large turgor pressures has a notable impact in the possibilities of growth and movement (see below).

Thus, plant cell walls form a dynamic intercellular structure that continuously remodels in response to physiological conditions. This structure plays a key role in homeostasis, growth, differentiation, defense against pathogens and, last but not least, movement, as we will later discuss in more detail.

5.5 Why Fungi and Plants Have Evolved Different Degrees of Constitutive Multicellular Complexity

A key characteristic of complex multicellularity is that not all cells have direct contact with the environment. This requires specialized cellular or intercellular mechanisms to transport oxygen and nutrients to inner cells. To address this challenge, complex organisms have evolved circulatory and respiratory structures—adaptations that are absent in simple cell aggregations or filamentous organisms. Based on this criterion, land plants are clearly complex multicellular organisms, whereas brown and red algae, as well as fungi, have achieved only limited levels of multicellular complexity.

We don't know when the ancestors of land plants achieved true multicellularity. Although evidence suggests that the last common ancestor of land plants and red and green algae possessed an ancient genetic toolkit for multicellularity—including molecular innovations related to signaling, environmental response, and growth—only plants, during their transition to land, achieved a qualitatively higher degree of multicellular complexity. Why? Research indicates that throughout the evolution of plants (and the broader clade of Embryophyta), genes involved in multicellularity were selectively favored because they enhanced the ability to adapt to environmental conditions and respond to stress. This developmental flexibility, combined with the ecological demands of terrestrial life, created opportunities for natural selection to favor genetic programs that operate in response to environmental cues.

As a result, plants evolved a highly plastic multicellular body capable of modulating growth in response to environmental changes. Such a growth program, intrinsically coupled to developmental processes, and subsequently shaped by the strong selective pressures operating in terrestrial habitats, was probably what enabled plants to reach higher levels of constitutive multicellular complexity.

5.6 Agency in Non-motile Multicellular Organisms

Given all the mentioned factors (sessile way of life and non-motility of the somatic cells), fungi, algae and plants act mainly through secretion, where all these different forms of non-motile multicellular organisms have developed extremely sophisticated strategies. Next, we will describe some of these sophisticated examples of secretory agency in fungi and plants. But in the case of plants (and some exceptional fungi), their higher constitutive complexity has also allowed other forms of agency, including macroscopic agency based on movement. These cases raise deep conceptual questions about their interpretation as true forms of agency because they are deeply entangled with constitutive-developmental processes (i.e., growth) constrained by tight biophysical limits.

Fungi The agency of multicellular fungi relies on detecting substances and secreting compounds, sometimes leading to remarkable predatory strategies. One striking example is *Ophiocordyceps unilateralis*, a parasitic fungus that manipulates the behavior of carpenter ants before ultimately killing them. The fungus begins as a spore that attaches to a passing ant. Once inside, it grows and alters the ant's behavior, compelling it to seek a warm, humid location ideal for fungal reproduction. As the infection progresses, the ant climbs a plant, locks its mandibles onto vegetation in a "death grip", and soon dies. The fungus then consumes the ant from within, eventually sprouting a stalk from its head that releases spores, continuing the cycle.

The manipulation is highly precise, as the fungus dictates the ant's final actions and selects attachment sites based on environmental conditions. Infected ants in temperate forests latch onto twigs, while those in rainforests attach to leaf veins. Sometimes, entire "graveyards" of infected ants are found, all positioned similarly for optimal spore dispersal. Research suggests that the fungus hijacks the ant's nervous system without directly invading its brain. Instead, it secretes neuromodulatory compounds that alter gene expression, affect muscle function, and interfere with chemical communication. Additionally, the fungus produces bacterial-like toxins that may disrupt social signaling within the colony. By suppressing stress responses and modifying neurotransmission, *O. unilateralis* ensures the ant's behavior aligns perfectly with its reproductive needs.

Interestingly, some multicellular fungi use specialized body structures combined with secreted substances to get their food. This way, they trap living nematodes, which then are slowly killed and decomposed. These body trapping structures include sharp spines (stephanocites, spiny balls, acanthocytes) and adhesive structures (webs, rings, and branching hyphae coated with sticky substances). A fibrillar layer of extracellular polymers aids in attaching traps to the nematode surface. Once a nematode is captured, the fungus forms a penetration tube that pierces the cuticle, leading to paralysis and rapid colonization by fungal hyphae. These nematophagous fungi can detect and respond to small molecules secreted by soil-dwelling nematodes, using these signals to initiate trap formation. Most fungal species trigger trap development only in the presence of prey, ensuring efficient energy use. Recognition and adhesion are the first crucial steps in nematode trapping, allowing these fungi to effectively hunt and consume their prey.

An interesting form of action in fungi is the explosive discharge of spores by the mucoromycota *Pilobolus* sporangium. This case may represent the physical limit of motion driven by hydrostatic pressure and elastic contraction to achieve high-speed spore dispersal (Fig. 5.2). This mechanism is a key feature of the asexual reproduction of this coprophilous fungus. The sporangium is propelled by a jet of cell sap, which deflects off the sporangium and extends significantly before breaking into

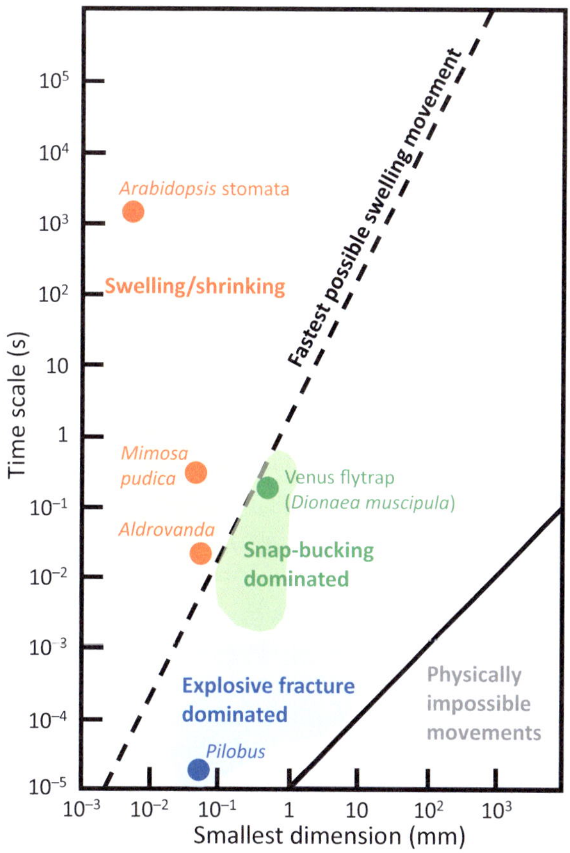

Fig. 5.2 Classification of fungal and plant movements. The duration of movement is plotted against the smallest dimension of the moving part. The lines represent performance limits, classifying movements into two categories: those constrained by fluid transport (above the dashed line) and those driven by elastic instabilities. The latter can be further divided into snap-buckling and explosive fracture. Both types of instabilities rely on geometries that gradually store elastic energy and release it suddenly. Examples include *Arabidopsis* stomata, *Mimosa* leaves, and the carnivorous plant *Aldrovanda* as swelling/shrinking-based movements, while its close relative, the Venus flytrap, exemplifies the snap-buckling mechanism. Finally, the fungus *Pilobolus*' sporangium discharge may represent the upper limit for explosive movements. Figure adapted from Skotheim and Mahadevan (2005)

droplets. Powered by hydrostatic pressure, this "squirt gun" mechanism is common among fungi in the Ascomycota and Zygomycota phyla. The turgor pressure driving the discharge is generated through osmotic processes, enabling sporangia to reach speeds of 2–25 m/s with accelerations between 20,000 and 180,000 *g*, propelling spores up to 2.5 m.

The process begins when the highly turgid subsporangial bulb ruptures along a weak line just below the junction between the sporangium and the columella. This rupture allows the elastic wall of the sporangiophore to contract, expelling a jet of sap that carries the sporangium forward. High-speed photography has confirmed this mechanism, capturing the stages of discharge and the role of the sap jet in spore propulsion.

A fascinating example of fungal agency is the symbiotic relationship in mycorrhizae, where both plant and fungal structures—with the likely participation of bacterial communities associated to the rhizosphere—undergo modifications to form a functional interface between plant roots and ectomycorrhizal hyphae. Mycorrhiza formation exemplifies fungal and plant agency, as it results from a complex exchange of secreted nutrients, along with immune and defensive responses, ultimately leading to mutualistic association. Before symbiosis begins, both organisms engage in a "prospective" biochemical dialogue, secreting compounds to initiate interaction. Each organism acts in its own interest, triggering and responding to signals from the other. The outcome is a mutually beneficial partnership that has evolved into an intimate association over time. In some cases, such as arbuscular mycorrhizae, the fungus penetrates the root cells of vascular plants, forming specialized structures called arbuscules—a shared tissue that enhances nutrient exchange.

Plants Among all sessile multicellular organisms, plants exhibit the most sophisticated forms of agency. Their defining characteristic is their highly plastic and dynamic multicellular organization, which enables continuous structural adjustments in response to environmental changes.

Land plants demonstrate a remarkable range of secretory actions as a primary means of agency. For example, tomato plants release volatile chemicals in response to caterpillar attacks, inducing the insects to fight each other while simultaneously warning nearby plants to activate defensive secretions. Some plants even engage in symbiotic cooperation with animals. A striking example is the African acacia tree, which has effectively "recruited" stinging ants for protection. These ants inhabit hollowed-out acacia thorns, feed on the tree's nectar, and aggressively defend it from herbivores and competing insects, ensuring the tree's survival. In addition to chemical communication aboveground, plants also secrete volatile compounds and exudates (liquids) belowground, establishing an exchange of information with other plants of the same or different species, sometimes also mediated by the mycorrhizal network.

Another essential aspect of plant agency is movement, a phenomenon that fascinated Charles Darwin, who extensively studied climbing and carnivorous plants, and various other forms of botanical motion. Plants exhibit a form of "constitutive adaptability" that blurs the distinction between growth and behavior. This adaptability often manifests in slow, movement-like responses driven by differential growth rates. Phototropism, for instance, allows plants to grow toward light, while gravitropism ensures that roots grow in response to gravity. Hydrotropism enables plants to adjust their growth toward moisture. A particularly striking example is the climbing bean (*Phaseolus vulgaris*), which wraps around a support in a slow, oscillatory motion known as circumnutation, a term introduced by Darwin and his son Francis in 1880. These nutational movements, regulated by proprioception, allow plants to control their posture in response to environmental stimuli.

Climbing species are widespread across the plant kingdom, appearing in nearly all angiosperm lineages and in select pteridophyte and gymnosperm families. Their climbing strategies display a surprising diversity and include twining stems, tendrils, hooks, and sticky adventitious roots. These mechanisms arise from modifications of different plant structures, including stems, leaves, and roots. Tendrils, for instance, have evolved independently multiple times, making them a notable case of convergent evolution. Despite early interest in tendrils since Darwin's time, the molecular mechanisms governing their development remain poorly understood, with only a few model systems, such as *Pisum* and *Vitis*, offering insights. Evidence suggests that diverse tendril formation involves distinct genetic regulatory networks, and some studies point to a connection between cytoskeletal changes and helical growth.

Beyond slow, adaptive growth "movements", some plants exhibit real forms of reversible and rapid motile behaviors. These movements are primarily driven by turgor pressure, a process also seen in fungi. Turgor-driven motions depend on the flow of water through cells, and their rate is constrained by how quickly water moves within plant tissues. The opening and closing of stomata, as well as the sleep movements of pulvini—specialized motor organs responsible for macroscopic leaf motion—are examples of such movements (Fig. 5.2). A classic case is *Mimosa pudica*, which folds its leaves when touched. This movement is controlled by the pulvinus, which regulates turgor pressure through changes in water flux. Interestingly, the genes involved in pulvinar movement also play broader roles in plant growth and development. The movement is initiated when sucrose is unloaded into intercellular spaces, altering water potential and triggering potassium ion efflux, leading to a sudden shift in turgor pressure.

A different mechanism governs the extremely rapid movements of the Venus flytrap, whose leaf closure is generated by a sudden osmotic pressure change. This process is facilitated by the transient opening of aquaporins, which allow water to rush in and trigger movement. Unlike growth-driven movements, the closure of the trap is fully reversible and depends on elastic instabilities and rapid cell-cell communication. The biophysical principles behind plant movements, such as turgor pressure and elasticity, impose strict limits on the speed and range of motion that plants can achieve (Fig. 5.2).

In summary, plant movements rely on diverse and complex mechanisms. Many of these movements are driven by changes in turgor pressure, often enhanced by elastic forces, which are essential for both slow and rapid movements. Water movement through plant tissues is fundamental, influencing both the speed and extent of movement. Some rapid movements result from mechanical instabilities, where stored elastic potential energy is rapidly converted into kinetic energy. Hormones, such as auxin—known for its role in regulating transmembrane proton flux and affecting cell turgor—along with electrical activity, play significant roles in linking stimulation to motor responses.

The existence of all these diverse adaptive behaviors raises an intriguing question: is it merely an internal adaptive reaction, or does it constitute a form of macroscopic motile action that actively shapes the environment? Perhaps both perspectives are valid. Circumnutation, for instance, is an adaptive growth response tightly integrated with plant development. However, reversible contractions linked to turgor variations resemble true motility, much like the extremely rapid leaf closure of the Venus flytrap. More advanced models are needed to distinguish between pure growth dynamics and mechanical tissue changes.

Ultimately, however, the organization of plants, constituted by fixed cells, imposes constraints on their possibilities to display complex movement-based behaviors. Their motility is ultimately governed by a rigid, interconnected cellular matrix that limits locomotive potential. From an evolutionary perspective, however, this is not a drawback but rather a structural feature that aligns with the plant kingdom's unique trajectory. The rigid cellular organization of plants has not hindered their complexity but has instead enabled the vast diversification of photosynthetic life, allowing them to thrive in an extraordinary range of ecological niches making plants the masters of sessile agency.[1]

5.7 Is Adaptive Growth a Form of Agency? Or the Essential Distinction Between Constitutive and Interactive Processes

An organism can adaptively modify its environment in two main ways: either by relying on its constitutive processes—those involved in its irreversible development and reproduction—or by using specialized mechanisms that enable reversible, interactive responses. Crucially, these adaptive responses must be modulated by sensory inputs to ensure they function effectively. Agency, therefore, requires flexible actions regulated by sensory control, typically organized as action-detection loops. For obvious reasons, responses based on constitutive processes tend to be slower, more energy-intensive, and less flexible than those mediated by reversible mechanisms.

[1] However astonishing plant agency may seem, we firmly distance ourselves from the terminological distortions that attribute anthropomorphic qualities and intelligence to plants—claims that often reflect subjective interpretations rather than empirical realities. Some have even gone so far as to propose a "plant neurobiology", citing Darwin's metaphor about a brain-like function at the tip of radicles. However, Darwin himself explicitly acknowledged in *The power of movement in plants* (1880, p. 572) that "plants do not of course possess nerves or a central nervous system."

This is why the majority of interactive behaviors typically identified as forms of agency—whether secretory or motile—rely on reversible systems that operate with a certain degree of independence from the organism's core constitutive machinery.

Nevertheless, as we have seen—both in the case of plants in this chapter and *Physarum* in the previous one—there are genuine examples of growth-based agency. In some instances, these growth-driven actions can even be repeated. Although developmental processes are generally based on irreversible mechanisms, this is not always the case in unicellular organisms. In plants, despite their complexity, growth and developmental processes are, to a large extent, reversible.

Yet, from the perspective of understanding the evolution and diversification of agency, growth-based forms represent a marginal case. This is because growth, like other constitutive processes in multicellular organisms, is generally too slow and energetically costly to support the kind of rapid, flexible, and repeatable interactions with the environment that characterize more complex forms of agency.

For this reason, the focus of this book has been primarily on secretory and motile actions supported by reversible mechanisms.

From our evolutionary perspective, a key feature of agency is the fact of being supported by reversible capable to evolve. Agency imply adaptive interactions—whether secretory or motile—which must be modulated by sensory inputs to ensure successful functionality. True agency, therefore, requires flexible actions driven by sensory control, thus structured as action-detection loops.

This is why, as discussed in the previous chapter, evolution imposes size-time constraints on motile agency: as an organism grows, maintaining flexible and rapid motility requires increasingly complex sensors, effectors, and information-processing systems. When resources are limited, trade-offs arise between motile capacity and other physiological demands, shaping evolutionary bifurcations in eukaryotic multicellularity.

Plants display highly flexible ontogenetic and growth processes that allow them to adapt dynamically to changing environmental conditions through slow developmental changes. For instance, plants can reallocate vascular resources based on the availability of light or moisture. In some cases, however, they are also capable of fast and reversible motile responses, which align more closely with the common-sense understanding of interactive agency. Behaviors such as the rapid closing of leaves when a carnivorous plant senses an insect clearly exemplify motile agency. Yet, this form of agency remains relatively rigid—limited to predefined reactions that do not evolve into more flexible or plastic behaviors. Such plasticity would require the modulation of action-detection loops, allowing the organism to adjust its actions based on varying conditions, a capacity that plants generally lack.

The reason lies in the special type of multicellular constitutive structure of plants. Karl J. Niklas and Stuart A. Newman have proposed that the unique multicellular architecture of plants results from an "active solid intercellular structure"—a stretchable and enzymatically modifiable extracellular matrix that differs fundamentally from the wall-less, motile multicellularity of animals (Fig. 5.3). They describe plant tissue as a "cellular solid", a composite material consisting of a solid phase (cell wall) and a liquid phase (e.g., cytoplasm or air).

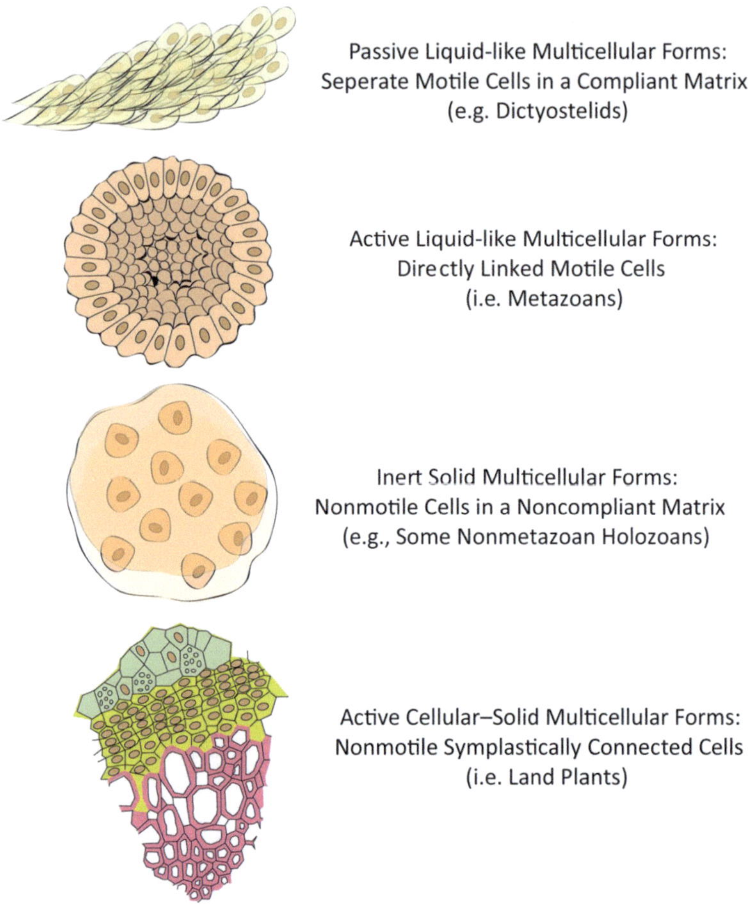

Fig. 5.3 Material properties and characteristics of tissues in multicellular organisms. Figure 1 from Niklas and Newman (2020), reproduced under the terms of the Creative Commons Attribution License (http://creativecommons.org/licenses/by/4.0/)

In this view, rapid and reversible macroscopic movements could be generated when plant intercellular structure functions as a hydrostatic system. Turgor pressure generates tensile stresses within cell walls, modulating tissue elasticity. As cell walls thicken or cytoplasm contracts, plant tissues transition from a hydrostatic state to a rigid solid, relying on the solid phase for mechanical stability rather than internal fluid pressure. This transition enabled plants to evolve supportive tissues, reducing their dependence on water for structural integrity and facilitating aerial growth—further reinforcing their sessile lifestyle.

From this perspective, land plant multicellularity led to the emergence of composite cellular solids, where tissues with distinct mechanical properties arise from variations in solid and fluid volume. Plants regulate these properties through metabolic control of cell wall composition and cytoplasmic viscoelasticity, allowing tissue mechanics to adapt to environmental conditions. Unlike animals, where cell migration enables processes such as wound healing and metastasis, the rigid adhesion between plant cells prevents cellular movement, precluding similar mechanisms.

While growth-based agency is not impossible—in fact, we acknowledge it in certain cases involving plants—it remains clearly marginal. This is because growth, like other constitutive processes, is typically too slow and energetically costly to support actions that require flexibility, speed, and reversibility in response to environmental changes. For this reason, in our discussion of the evolution and complexification of agency, we have limited the concept to the domain of *adaptable, reversible actions*, which primarily includes secretory and motile behaviors.

Comparing Plant and Animal Movement To conclude this chapter and start advancing the issues we will deal with in the next, let us summarize key analogies and differences between plant and animal movements. Animals possess well-developed sensory and locomotor systems that allow them to move freely and avoid environmental stresses. In contrast, plants are sessile and rely on growth movements, such as circumnutation, to interact with their surroundings. While animal movement is dynamic and often involves whole-body locomotion, plant movement is typically limited to specific organs or parts.

The underlying mechanisms of movement also differ significantly. Animal movement is driven by muscle (i.e., actomyosin filaments) contractions coordinated by the nervous system, which require frequent sensory information updates. Instead, plant movement relies on changes in turgor pressure (i.e., osmotic processes governed by water flow) and mechanical instabilities. Animal locomotion exhibits diverse patterns varying within and among individuals and populations. In contrast, plant movement is more uniform, often involving repetitive growth patterns like the helical motion of tendrils.

In plants, development itself structures both body formation and environmental adaptation. Consequently, their motile adaptive actions are less rapid and lack the versatility and speed of motile organisms. For instance, when analyzing adaptive bending, researchers emphasize that auxin, a hormone governing growth asymmetry, integrates mechanochemical feedback loops to shape plant morphology. However, this bending also influences morphogenesis, creating a complex interplay between development and environmental response.

Furthermore, the functional integration, homeostasis, and cohesion of animal body plans rely critically on the migration and patrolling behaviors of certain cell types, such as immune, connective, and repair cells. The way plants and fungi address challenges related to distinguishing themselves from their environment—and defining, maintaining, and caring for their individuality—differs significantly from the animal approach. In animals, many essential functions, such as eliminating pathogens or clearing debris, are carried out by cells that can migrate and perform actions like phagocytosis.

In summary, the evolutionary splits that led to radically different ecological niches—sessile autotrophy in plants and sessile heterotrophy in fungi—favored the development of multicellular forms based on cell wall interactions and associations.

This imposed a fundamental constraint on the types of body plans that could evolve later, particularly limiting the diversification and innovation of multicellular agency. In contrast, a small group of protists without cell walls evolved into contractile, multicellular forms with flexible and rapid motility.

In the next chapter, we will explore how metazoans evolved a solution that supports the open-ended complexification of interactive behaviors—particularly fast and flexible multicellular motility—without undermining the complexity of developmental and physiological organization. This solution is grounded in the relatively high degree of freedom that individual eukaryotic cells retain to move within the body, allowing them to control the identity of neighboring cells, form dynamic connections, and exert mechanical forces to maintain and adapt the organism's integrity.

Bibliography

Arnellos A, Moreno A (2015) Multicellular agency: an organizational view. Biol Philos 30:333–357. https://doi.org/10.1007/s10539-015-9484-0

Bich L, Pradeu T, Moreau JF (2019) Understanding multicellularity: the functional organization of the intercellular space. Front Physiol 10:1170. https://doi.org/10.3389/fphys.2019.01170

Bowles AMC, Bechtold U, Paps J (2020) The origin of land plants is rooted in two bursts of genomic novelty. Curr Biol 30:530–536.e2. https://doi.org/10.1016/j.cub.2019.11.090

Bringloe TT, Starko S, Wade RM et al (2020) Phylogeny and evolution of the brown algae. Crit Rev Plant Sci 39:281–321. https://doi.org/10.1080/07352689.2020.1787679

Darwin C [assisted by F Darwin] (1880) The Movements of Plants. John Murray, London

Feng X, Zheng J, Irisarri I et al (2024) Genomes of multicellular algal sisters to land plants illuminate signaling network evolution. Nat Genet 56:1018–1031. https://doi.org/10.1038/s41588-024-01737-3

Fritz-Laylin LK, Lord SJ, Mullins RD (2017) Our evolving view of cell motility. Cell Cycle 16:1735–1736. https://doi.org/10.1080/15384101.2017.1360655

Grosberg RK, Strathmann RR (2007) The evolution of multicellularity: a minor major transition? Annu Rev Ecol Evol Syst 38:621–654. https://doi.org/10.1146/annurev.ecolsys.36.102403.114735

Hutchings MJ, de Kroon H (1994) Foraging in plants: the role of morphological plasticity in resource acquisition. Adv Ecol Res 25:159–238

Jiao C, Sørensen I, Sun X et al (2020) The Penium margaritaceum genome: hallmarks of the origins of land plants. Cell 181:1097–1111. https://doi.org/10.1016/j.cell.2020.04.019

Jonsson K, Ma Y, Routier-Kierzkowska AL, Bhalerao RP (2023) Multiple mechanisms behind plant bending. Nat Plants 9:13–21. https://doi.org/10.1038/s41477-022-01310-y

Kingsland SE, Taiz L (2025) Plant "intelligence" and the misuse of historical sources as evidence. Protoplasma 262:223–246. https://doi.org/10.1007/s00709-024-01988-1

Knoll AH (2011) The multiple origins of complex multicellularity. Annu Rev Earth Planet Sci 39:217–239. https://doi.org/10.1146/annurev.earth.031208.100209

Knoll AH, Hewitt (2011) Phylogenetic, functional, and geological perspectives on complex multicellularity. In: Brett Calcott B, Sterelny K (eds) The major transitions in evolution revisited. MIT Press, Cambridge, MA. Online edn: MIT Press Scholarship Online https://doi.org/10.7551/mitpress/9780262015240.003.0013

Miao L, Yin Z, Knoll AH et al (2024) 1.63-billion-year-old multicellular eukaryotes from the Chuanlinggou Formation in North China. Sci Adv 10:eadk3208. https://doi.org/10.1126/sciadv.adk3208

Money NP (2023) The fastest short jump in nature: progress in understanding the mechanism of ballistospore discharge. Fungal Biol 127:835–844. https://doi.org/10.1016/j.funbio.2023.01.001

Nagy LG, Varga T, Csernetics Á et al (2020) Fungi took a unique evolutionary route to multicellularity: seven key challenges for fungal multicellular life. Fung Biol Rev 34:151–169. https://doi.org/10.1016/j.fbr.2020.07.002

Niklas KJ, Newman SA (2020) The many roads to and from multicellularity. J Exp Bot 71:3247–3253. https://doi.org/10.1093/jxb/erz547

Skotheim JM, Mahadevan L (2005) Physical limits and design principles for plant and fungal movements. Science 308:1308–1310. https://doi.org/10.1126/science.1107976

Yafetto L, Carroll L, Cui Y et al (2008) The fastest flights in nature: high-speed spore discharge mechanisms among fungi. PLoS ONE 3:e3237. https://doi.org/10.1371/journal.pone.0003237

Chapter 6
Metazoans: An Explosion of Agency

Abstract This chapter explores the evolutionary emergence and significance of metazoans, emphasizing their pivotal role in the expansion of biological complexity. It examines how the transition from unicellular to multicellular life introduced radically new body organizations that enabled novel forms of motility and more dynamic interactions with the environment. These new organizational structures also made possible the emergence and evolution of nervous systems, which in turn supported coordinated movement, sensory integration, and increased behavioral complexity. The chapter traces the origins of neural networks and analyzes the interplay between body plans and neural development. It highlights how enhanced motility and sensorimotor integration led to the emergence of the most sophisticated forms of agency in the biological world. Ultimately, the discussion underscores the reciprocal relationship between brain development, bodily organization, and ecological adaptation—factors that have collectively shaped the evolutionary trajectory of animals.

Keywords Metazoan evolution · Nervous system origins · Cephalization · Brain-body integration

6.1 Origin and Significance of Metazoans

The emergence of metazoans (or animals), with their remarkable diversification and complex body plans, is one of the most significant evolutionary events in life's history. Their appearance, which has occurred only once in the history of life, introduced a fundamentally new type of multicellular, heterotrophic organism capable of sensing and acting as a unified whole while exhibiting an immense variety of moving behaviors, both at the global organismic level and within it (e.g. cell migration). In turn, this innovation created an entirely new and highly diversified ecological domain.

Á. Moreno, J. Peretó, *An Evolutionary Story of Agency*, Foundations for Interdisciplinarity in the Life Sciences: Concise Monographs,
https://doi.org/10.1007/978-3-032-05644-3_6

6.1.1 Origins of Metazoans

Although there is less consensus on the origin of metazoans than on the emergence of eukaryotic cells, their appearance represents one of the most profound evolutionary transitions in life's history. Metazoans belong to the broader Holozoa group, which includes various unicellular protists (e.g., choanoflagellates, ichthyosporeans, and filastereans) that evolved from the Opisthokonta supergroup over a billion years ago. This supergroup also gave rise to fungi (see Chap. 5). Unlike many protists with protective cell walls, the unicellular ancestors of metazoans lacked rigid walls, allowing for amoeboid contractility—an essential feature for the evolution of complex multicellularity.

Unicellular holozoans, the direct ancestors of metazoans, display significant diversity but share key developmental processes and genetic toolkits with animals. Many genes essential for multicellular functions, such as those involved in cell-cell adhesion and extracellular matrix interactions, were already present in these ancestors and repurposed in early metazoans. However, despite the intense researches, how the first obligate multicellular metazoan appeared from these unicellular ancestors is still, as Ruiz-Trillo and colleagues have pointed out, an enigmatic and unpredictable event with far-reaching consequences (Fig. 6.1).

The transition likely began 750–800 million years ago in an aquatic environment, and could have occurred through different mechanisms: clonal division within colonial-like choanoflagellates, multicellular development from a multinucleate coenocyte, or aggregation-based multicellularity. Some unicellular relatives of animals, such as *Chromosphaera perkinsii*, exhibit prolonged multicellular stages with coexisting distinct cell types, offering potential models for early metazoan evolution. However, the limited diversity among surviving unicellular relatives complicates reconstructions of this transition.

Despite these uncertainties, all metazoans share a single evolutionary origin. The absence of intermediate stages suggests that once this kind of multicellularity emerged, it followed a highly constrained evolutionary trajectory, likely due to collective constraints on maintaining viability. Evidence indicates that the unicellular ancestor of metazoans had a complex life cycle, forming multicellular structures through clonal division, aggregation, or both. This lineage displayed an early tendency to innovate in genes related to multicellularity, a trend that continued in animals.

The first metazoans were heterotrophic multicellular aquatic organisms capable of regulating cell differentiation via hierarchical gene regulatory networks, enabling spatial organization of cells. This transition led to a new developmental paradigm characterized by collective cellular reshaping, irreversible remodeling, cell migration, and key signaling pathways controlling cell fate. These innovations allowed early animals to exhibit coordinated, reversible behaviors—crucial for both their development and their ability to explore new forms of motility. In the following sections, we will explore the significance of this transformation in greater depth.

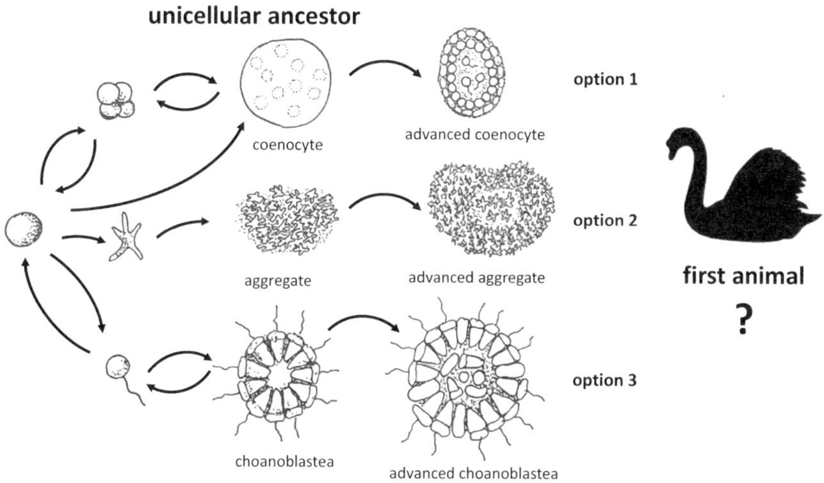

Fig. 6.1 Three possible scenarios for the origin of animals as discussed by Ruiz-Trillo et al. (2023). The unicellular ancestor of animals likely had a complex life cycle with multiple potential pathways to temporary multicellularity. The transition to the first true animal was probably a rare and unpredictable "black swan" event, making it difficult to reconstruct due to incomplete taxonomic, historical, and biological data. Given these uncertainties, it is essential to remain open to different explanations for the origin of animals. There are three possible evolutionary paths from this ancestor. One suggests that animals originated from a coenocytic stage, similar to extant ichthyosporeans. Another posits that multicellularity arose through cell aggregation. The third, combining elements of the choanoblastaea and synzoospore theories, proposes that animals evolved from a colonial stage resembling modern choanoflagellates. While these scenarios are all plausible, the possibility of alternative, yet undiscovered, pathways should not be overlooked. Adapted from Fig. 4 by Ruiz-Trillo et al. (2023). Black swan silhouette reproduced under a Creative Commons License from Phylopic (https://www.phylopic.org/)

6.1.2 *The Role of Motility in the Origin of Metazoans*

One of the most defining features of the transition to the first multicellular metazoans was the transformation of the single-cell motility machinery into an integrated, multicellular contractile system. This system enabled coordinated contractility during development and, crucially, maintained somatic cell motility throughout the life cycle, allowing for a number of fundamental functions to be carried out by migrating cells (e.g., connective, immune or housekeeping functions), on top of managing flexible movement at the organismal level. The evolution of this new set of capabilities likely made the transition to multicellularity particularly complex and challenging.

The animal body is inherently soft and flexible, formed by epithelia that fold and grow during development. Unlike plants and fungi, animal cells lack rigid cell walls, making them structurally dynamic. While some animals have evolved specialized skeletal elements for support, these adaptations remain secondary and are not universally present. Despite these reinforcements, the fundamental organization of animals is highly malleable, with muscle cells (or their precursors) enabling rapid shape changes for stabilization and movement. This ability to contract and reshape the body is essential for initiating and controlling motion, setting animals apart from other multicellular life forms.

In addition to contractility, a fundamental feature of metazoan body organization is the internal migration of cells. Migratory phagocytic cells gradually evolved into specialized types, capable not only of recognizing and eliminating bacteria but also of clearing dead cells—thus playing a vital role in maintaining global homeostasis during both embryonic development and adult life. The phylogenetic distribution of macrophage-like cells suggests that they emerged early in metazoan evolution. For example, in Placozoa, fiber cells are capable of phagocytosing cell debris and bacteria and also participate in wound healing (see below).

Metazoans have evolved a uniquely integrated body capable of performing mechanical work—changing shape and actively moving. This intrinsic ability for coordinated multicellular contractility has led to the evolution of a vast diversity of rapid, reversible macroscopic movements, opening new ecological opportunities and expanding the range of available niches.

However, developing coordinated multicellular motility requires a high degree of intercellular integration. The evolutionary expansion of animal movement is rooted in the contractility of individual cells and their capacity for coordination. This innovation was achieved by extending the intracellular actomyosin complex—responsible for cell growth, reproduction, and motility—into a supracellular system (Fig. 6.2). This intercellular structure enabled the wall-less cells of ancestral holozoan protists to coordinate actomyosin assembly across multiple cells, generating contractile networks through actomyosin supracellular cables.

This transformation created a new type of intercellular environment, which Stuart Newman describes as a "liquid-like multicellular active material" (see last section, Chap. 5) where cellular motility allows constant rearrangement of neighboring cells. Intercellular attachments provide flexible, fluid cohesion for contractile tissues while also permitting controlled cellular movement within the tissue. Together, these mechanisms maintain tissue integrity, cell migration and homeostasis while enabling large-scale morphogenesis and coordinated movement. This supracellular motor system has allowed metazoans to achieve both irreversible developmental transformations and reversible behavioral movements. As a result, animals have been able to develop a diverse and adaptable motile-based agency. To fully understand this profound evolutionary shift, we must explore in greater detail how it might have occurred.

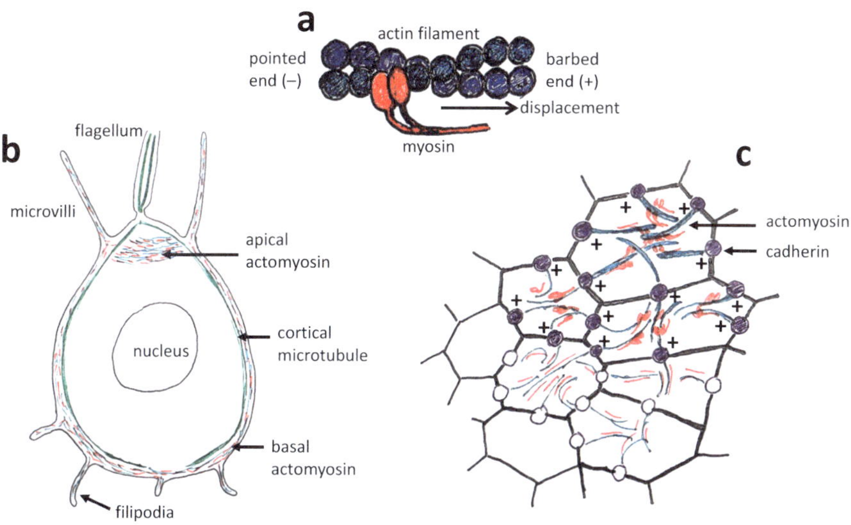

Fig. 6.2 (**a**) Actin filaments (blue) show polarity from pointed end (–) to barbed end (+) and can interact with myosin-II motor (red) that can drive displacement using the energy of ATP. (**b**) Choanoflagelates contain diverse cytoskeleton elements: microtubules (green) and actomyosin complexes (blue and red) at the basal and apical poles, underneath filopodia and microvilli, respectively. (**c**) Illustration of a tissue composed by cells linked at the junctions (cadherin in violet) that can interact through contractile actomyosin fibers and deform the tissue. Figure based on ideas discussed by Brunet (2023)

6.1.3 How Could the Transition to Multicellular Motility Have Taken Place?

The foundation of this major evolutionary shift, as Thibaut Brunet argues, was the transformation of the actin/myosin II (actomyosin) complex to organize two distinct yet functionally integrated systems of intercellular contractility. One system supports slow, irreversible contractile movements that drive the formation of epithelial and mesenchymal tissues, shaping the development and morphogenesis of all metazoans. The other enables fast, reversible contractions, leading to the emergence of muscle tissues that generate macroscopic motile behavior.

The actomyosin complex, which dates back to the last common ancestor of amoebozoans and animals, plays a versatile role in metazoan development, including morphogenesis, cell migration, and epithelial maintenance. Even in unicellular organisms, actomyosin supports two fundamentally different types of deformation: irreversible (plastic) changes, such as actomyosin ring contraction during cytokinesis, and reversible (elastic) deformations seen in crawling cells. But the emergence of metazoan multicellularity required integrating these two contractile modes into a single, coordinated framework.

Two major innovations marked this transition. (1) Intercellular coordination: the ability to integrate actomyosin activity across multiple cells, forming supracellular cables that maintain tissue integrity while enabling large-scale morphogenesis and movement. (2) Specialization of contractile cell types: the evolution of two functionally distinct cell types: one specialized in slow, sustained contractions (resembling smooth muscle and non-muscle contractile cells), and another specialized in fast contractions (precursors of striated muscle). Initially, both cell types may have originated from generic contractile epithelial or mesenchymal cells, later diverging to support either morphogenetic remodeling or behavioral movement. Some cells even transformed almost entirely into actomyosin-based structures, leading to the emergence of muscle cells.

Together with these intercellular developments, key intracellular transformations were also necessary. One of the most critical transformations was apical constriction, where contraction of a cell's apical side changes its shape, driving epithelial invagination and other morphogenetic processes. Coordinated apical constriction of multiple cells generates curvature changes in epithelial layers, essential for tissue folding and organ formation. However, maintaining structural balance required precise control of membrane elasticity and endocytosis, which prevented excessive tension that could hinder invagination. Pulsed actomyosin contractility, combined with endocytic remodeling, acted as a ratcheting mechanism, ensuring progressive, irreversible shape changes.

Interestingly, primitive forms of apical constriction already exist in choanoflagellates. In *Choanoeca flexa*, a colonial choanoflagellate, actomyosin is distributed in both apical microvilli and basal filopodia, with myosin II concentrated in these regions. In response to darkness, a light-sensitive protein triggers coordinated actomyosin contraction, causing colony-wide inversion. This transition switches the colony from a feeding conformation (flagella facing inward) to a swimming/escape mode (flagella outward), illustrating an early example of collective contractility governing macroscopic behavior.

Crucially, for this multicellular contractile system to interact effectively with its environment, it had to develop mechanisms for coordinating cell movement in response to stimuli. In this regard, the role of voltage-gated calcium channels in intercellular electrical signaling for synchronizing activity was just as important as the contractile machinery itself.

Although primitive forms of voltage-gated calcium channels are found in many protists, these ancestral voltage-sensitive signaling mechanisms did not support the regulation of cytoplasmic protein-protein interactions involved in subcellular localization and functional complex formation. In contrast, metazoan voltage-gated calcium channels evolved to enable finely tuned control over these interactions, allowing precise coordination of subcellular organization and complex cellular functions.

Recent discoveries in the colonial choanoflagellate *Salpingoeca rosetta* reveal that voltage-gated calcium channels involved in calcium-mediated electrical signaling orchestrate both asynchronous and synchronous cellular events, suggesting early forms of regulated communication between cells. Notably, *S. rosetta* exhibits ciliary arrest coupled with apical-basal contraction, a behavior synchronized across the colony, allowing coordinated contraction in response to environmental changes. This suggests that even before the emergence of true multicellularity, actomyosin-based contractile systems had already begun functioning as collective sensory structures, capable of dynamic adaptation.

In sum, this fundamental contractile organization paved the way for the evolution of diverse, rapid, and reversible forms of macroscopic motility. Even in its incipient form among choanoflagellates, this innovation had profound consequences. It established the structural foundation for increasingly complex motile behaviors at a multicellular scale, ultimately enabling the vast locomotor diversity seen in metazoans.

However, key questions remain: How did the first intercellular actomyosin cables emerge? How was contractile function divided between fast and slow myosin? Since myosin I and II first diverged in ancient unicellular ancestors, they may have initially been part of distinct contractile machineries located in different cell regions or life-cycle phases.

In the early stages of metazoan evolution, contractile tissues likely played dual roles in both morphogenesis and movement. The transition between plastic and elastic deformations in actomyosin networks may not have been mechanically complex. Pulsed actomyosin contractions naturally exhibit elastic relaxation, but irreversible tissue remodeling requires additional 'ratcheting' mechanisms, such as endocytic remodeling of cell junctions. The evolutionary gain or loss of such ratcheting systems could have determined whether an actomyosin network primarily supported elastic (reversible) or plastic (irreversible) deformations.

All this process could have been driven by some pre-existing genes and structures, originally serving unicellular functions, that could have been repurposed for morphogenesis simply due to the new spatial and mechanical constraints of multicellularity. As cell clusters increased in scale, novel mechanical forces and self-organizing physical processes could have emerged, creating new developmental contexts in which existing genetic toolkits acquired new functions. In this sense, the transition to multicellularity was largely a process of self-organization.

Nevertheless, even if self-organization may explain certain important aspects of this transition, scaling alone was not enough. The evolution of multicellular contractility also required new genes and specific molecular machinery to drive the formation of intercellular junctions, ensuring coordinated contractions across tissues. Additionally, cell-cell communication mechanisms—mechanical, chemical, and electrical—were essential for synchronizing these movements across the developing organism.

The construction of this radically new flexible architecture, capable of contractile movements, had far-reaching consequences in the history of agency because it enabled both an extraordinary expansion of motile behavior and a vast increase in developmental and physiological complexity. Unlike plants and fungi, which rely on rigid cell walls, metazoans evolved a dynamic intercellular structure where individual wall-less cells deform collectively, forming an integrated contractile system.

This flexible body plan not only supports large-scale coordinated movement but also underpins the sophisticated interplay between morphogenesis and behavior.

This evolutionary innovation provided an elegant solution to a fundamental challenge: balancing the developmental and physiological demands of an increasingly complex multicellular organism with the need for unrestricted movement and behavioral adaptability. Remarkably, both developmental morphogenesis and active movement emerged from the same fundamental structure—an intercellular actomyosin network—allowing for a seamlessly integrated system where two functionally distinct yet structurally intertwined machineries evolved in parallel.

6.1.4 Metazoan First Steps

Thanks to this new organization, early metazoans emerged as the first true multicellular organisms capable of constructing a flexible body—primarily through epithelial tissues—that could function as a unified, free-moving entity, capable of sensing macroscopic properties. Their key features included a multicellular structure that created an internal domain distinct from the external environment, contractile surface apparatuses sensitive to tension, and body contraction as a primary mode of movement rather than motility itself.

It is worth considering how such a highly specific form of multicellular organization could exhibit functional behaviors that, in turn, shaped its evolutionary trajectory. While the details remain uncertain, as proposed by Nguyen and Farge, morphogenetic processes in early metazoans were probably coupled to their active feeding behavior. Consequently, selective pressures may have favored gastrulating structures initiated by marine flow, which enhanced nutrient uptake compared to inert forms. These early multicellular assemblies likely preceded primitive animals, with mechanotransduction processes conserved as essential components of modern gastrulation.

A fundamental aspect of early animal evolution was the ability to link sensation and action, using available resources to establish functional connections between them. The tensile components of the early animals could change length through contraction and relaxation, making their tensegrity-based structure dynamic and adaptable. These changing mechanical forces not only shaped the body's structure but also served as internal signals, traveling across large regions—much like a rope pulling a distant bell—and influencing biochemical processes at the cellular level. Mechanosensitive cellular sensors responded to these deformation or contraction stresses, generating reafferent signals—a crucial feedback mechanism. The body's architecture and sensory systems coevolved to exploit this capacity, integrating mechanical feedback into an increasingly sophisticated control system.

A pivotal innovation in metazoan evolution was the emergence of neurons. While neurons represented a groundbreaking development with far-reaching consequences, their success was largely enabled by the pre-existing multicellular organization based on contractile epithelial tissues. These tissues provided the functional foundation for early nervous systems, facilitating both coordination and communication across the body.

Neurons were transformative for several reasons. First, they enabled the development of nervous systems, establishing interconnected communication pathways. Also enhanced the ability to process macroscopic sensory information, allowing for more effective interaction with the surroundings. By integrating different types of sensory input, neurons facilitated increasingly diverse and efficient behavioral responses. This is why, with the advent of nervous systems, early metazoans gained the ability to execute more complex and adaptive actions, linking their internal structural complexity to behavioral success, which allowed animals to conquer new ecological niches.

In the following sections, we will explore how primitive metazoans transformed some of their cells into early neurons, their original functions, and how these neurons evolved to take on new roles. This evolutionary trajectory—the prehistory of the nervous system—ultimately led to the emergence of animals with behavioral capacities and an internal complexity in which the nervous system became central.

6.2 Origins and Consequences of the Nervous Systems

6.2.1 Origins of the Nervous Systems

It is widely accepted that the last common ancestor of early animals likely had some primitive form of epithelial tissue but lacked neurons. Modern animals like sponges and placozoans, which also lack neurons, can still perform essential behaviors for their survival. Sponges are marine sessile but exhibit various contractions to filter food, while placozoans crawl using coordinated ciliary movements to locate food. In both cases, coordinated motility relies on primitive epithelial tissue and the coordination of cell movements through chemical and electrical signals. Under certain conditions, such as when searching for food, all cells within the organism can move together in a coordinated manner.

Placozoa are very small animals, which move in marine soil through a coordinated cilia movement. This collective behavior is governed by paracrine signaling (i.e., cell signals affecting other nearby cells), with small neuropeptides secreted by peptidergic cells lacking synapses or cellular projections. For example, *Trichoplax adherens* uses neurotransmitters such as epinephrine to control its negative taxis behavior, with ligand-mediated triggering of calcium signals that modulate ciliary redox states. These signals ultimately influence cilia movement, coordinating the organism's motion. Additionally, epithelial cells exhibit rapid contractility, aiding in locomotion and feeding. This suggests that, with the emergence of multicellular animals, small neurotransmitters evolved to facilitate cell-to-cell communication.

Perhaps even more intriguing is that one of the six identified cell types in *Placozoa*—the fiber cells—are capable of migration, phagocytosis, wound healing, and possibly even coordinating movement. These functions suggest an early form of immune activity. Phagocytosis carried out by migratory cells appears to be one of the earliest metazoan mechanisms for maintaining organismal identity, performing housekeeping functions, and preserving homeostasis.

Although placozoans lack conventional neural systems, they employ unique signaling pathways to synchronize cellular movements. Fibrillary cells in *Trichoplax* secrete adrenergic signals and regulate cell coordination via specialized membrane receptors and long filopodia that connect neighboring cells. This cell type, with multiple axonal processes, is crucial for enabling multicellular organisms to sense external signals and modulate intercellular communication, marking a significant evolutionary development.

However, chemical signaling has limitations in larger organisms, as the diffusion of chemical signals becomes less efficient with increasing body size, reducing coordination. This raises the question of how pre-neural mechanisms evolved into the first nervous systems in animals.

Key neuronal gene modules found in non-neuronal secretory cells of placozoans suggest that neurons could have evolved from such cells, as proposed by the "chemical brain" hypothesis. According to this hypothesis, early nervous systems may have evolved from systems that coordinated multicellular motility through the release of signaling molecules that affected target cells expressing specific receptors. Early synapses likely connected multiple sensory-neurosecretory cells into networks, enabling synchronized peptide release, which improved coordination and effector responses. This suggests that the key building blocks of the nervous system were present early in metazoan evolution, even before the development of fully functional neurons.

Tracing the phylogenetic origin of nervous systems in metazoans is challenging, with debate over what qualifies as a neuron. While neurons of animals with bilateral symmetry (i.e., bilaterians, see Sect. 6.2.3) are highly polarized cells with dendrites, axons, and synapses, primitive neurons resemble those found in ctenophores—sister group to all animals with nervous systems. These ctenophores have a subepithelial nerve net with neuronal cells interconnected via continuous plasma membranes, without synapses. Despite lacking long cell projections (i.e., neurites) and synapses, ctenophores exhibit diverse behaviors, such as controlling water flow for movement and feeding.

The next critical step in nervous system evolution was the development of polarized neurons capable of forming long membrane projections that could connect through electrical impulses and chemical signals (Fig. 6.3). This transition marked a pivotal moment in metazoan evolution, enhancing motility. Although the exact timing and location of this development remain unclear, the neural tracts found in modern cnidarians suggest that the evolution of axon guiding and fasciculation mechanisms was key to forming large networks of neurons capable of coordinating complex behaviors, such as visually guided swimming in cubozoan medusae and associative learning.

The emergence of the nervous system had profound implications for metazoan evolution. To understand its significance, we must first examine its operational

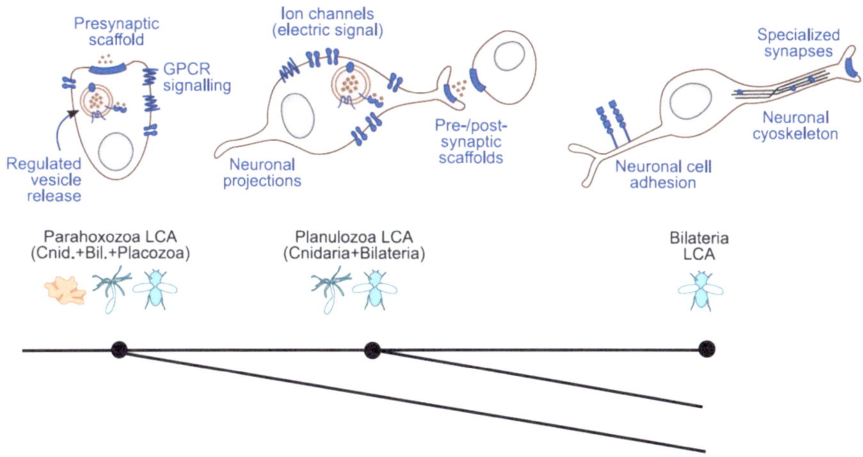

Fig. 6.3 Neurons originated from primitive fibrillary peptidergic cells (left), similar to those found in extant placozoa. These cells secrete adrenergic signals and regulate coordination via G-protein coupled membrane receptors (GPCR). Early neurons interconnected without synapses or axodendritic polarity were likely present in the last common ancestor (LCA) of Cnidaria and Bilateria (center). Finally, polarized neurons with dendrites, axons, and specialized synapses were active in the LCA of Bilateria (right). Adapted from Najle et al. (2023)

nature and how it differs from other physiological systems in metazoans, which made it a fundamentally different and revolutionary development.

6.2.2 The Organizational Specificity of the Nervous System and Its Implications

Why has the emergence of neurons and synaptic intercellular communication been so crucial in the evolution of metazoan behavior? And why has this evolution followed such diverse paths? To answer these questions, we must first understand the qualitative novelty introduced by the nervous system.

The nervous system consists of electrically excitable neurons, specialized cells with adaptable projections that enable complex connectivity. Although originally derived from secretory cells, neurons evolved a unique electrochemical communication system based on action potentials, allowing for two key phenomena. First, by converting diverse physical inputs into neural spikes, they establish a universal informational currency, facilitating integration and learning. Second, their electrochemical conductivity, combined with the networked architecture of the nervous system and the influence of neuromodulators, creates a dynamic, high-dimensional, nonlinear, and recursive domain. Neuronal spikes regulate their own activity through modulatory feedback, generating a self-organizing system in which all units continuously interact and influence one another.

Functionally, the nervous system operates as a dynamic network where every element contributes to the overall activity. Even in the absence of external stimuli, it generates intrinsic neurodynamic patterns. Without continuous neuronal interactions, the system's structural integrity would deteriorate, as many synaptic connections depend on ongoing activity for their maintenance. This makes the nervous system an unparalleled form of intercellular organization, capable of processing vast amounts of information at high speed and in parallel. Unlike any other biological system, it integrates both external and internal stimuli to coordinate complex actions efficiently.

The nervous system represents an unprecedented biological innovation. It enables the rapid generation and modification of countless informational configurations, making it a uniquely powerful regulatory system. Unlike other physiological processes, which primarily respond to environmental or internal signals, the nervous system operates autonomously, regulating not only interactions with the environment but also internal physiological and developmental processes.

One of its most transformative effects has been the expansion of behavioral complexity. By enabling more advanced motile and sensory organs, as well as sophisticated sensory processing, the nervous system has continuously shaped the evolution of animal behavior. Moreover, successful behaviors reinforce or eliminate neural connections, creating a feedback loop between neuronal evolution and agency. This interplay between behavior and neural adaptation has played a fundamental role in the co-evolution of nervous systems and animal agency.

When examining the role of nervous system complexification in the evolution of metazoan agency, we focus on the internal mechanisms that govern external behavior. In this context, learning—especially associative learning—emerges as a key driver in the development of sensory and motor capacities.

Learning is a fundamental capacity for successful action because it modifies the neural processes that guide behavior, for example by creating new associations between sensory inputs and motor outputs. From a neurophysiological perspective, the evolution of learning mechanisms is closely tied to the increasing complexity of the nervous system, particularly the emergence of integrated control centers such as the brain. However, even in the decentralized nervous systems of cnidarians, we can already observe basic forms of associative learning.

Beyond behavior, the nervous system also regulates crucial developmental and physiological processes. It controls growth, metamorphosis, molting, and regeneration through hormonal signaling, while also overseeing vital functions such as circadian and circalunar rhythms, immunity, metabolism, digestion, and diuresis. In some cases, such as feeding and gut movement in sea anemones, the distinction between behavior and physiology becomes blurred.

These two regulatory roles—behavioral and physiological—are deeply interconnected. Complex behavioral success requires finely tuned body regulation, just as the ability to develop and sustain a complex body is essential for the evolution of increasingly sophisticated nervous systems.

Thus, by enabling increasingly complex motile behavior, the nervous system introduced a profound evolutionary shift. From its earliest stages, interconnected neural networks assumed control over the coordination of the entire metazoan body. This led to a mutual dependency: neural control facilitated the emergence and maintenance of more complex body structures, while these structures, in turn, became increasingly reliant on the nervous system and the interactive loops that this enables with the changeful external world. Moreover, the nervous system itself depended on the success of the behaviors it orchestrated. This interdependence reshaped body development and function, making the nervous system a powerful driver of evolutionary innovation.

The emergence of the nervous system marked a turning point in metazoan evolution, introducing an entirely new mode of body construction. With its ability to impose an open-ended array of regulatory controls on physiological and developmental processes, it fundamentally altered the trajectory of evolution.

However, several factors have constrained the evolution of nervous system complexity across different lineages. Neurons are energetically costly, as are the sensory and motor organs they coordinate, and their level of sophistication is closely tied to the organism's overall body plan. In bilaterians, the emergence of glial cells introduced additional layers of developmental and homeostatic regulation within the nervous system. Among these, microglia and astrocytes play key roles in the central nervous system's immune functions. They continuously monitor for pathogens and initiate immune responses, including phagocytosis. These non-neuronal cells have come to play increasingly important roles in the maintenance, regulation, and complexification of nervous system function.

The extent to which a nervous system can increase in complexity is therefore limited by the structural framework within which it operates. Consequently, nervous system evolution, including trends toward cephalization, has always been closely linked to changes in body architecture that either enable or necessitate such complexity.

Ultimately, the history of animal evolution has been shaped by a dynamic interplay: increasing nervous system complexity has driven and amplified body plan diversification, while the constraints and opportunities presented by body architecture have, in turn, influenced nervous system evolution. To fully understand this process, we must now examine how the potential of the nervous system has been realized, exploring the different evolutionary pathways it has taken and the consequences of these divergences.

6.2.3 Early Steps in the Evolution of the Nervous Systems

Although the exact body plans of early metazoans remain unknown, it is likely that both digestion and circulation occurred within a space similar to the gastrovascular cavity found in modern cnidarians. In these organisms, the nervous system is essentially a diffuse net, where sensory and ganglionic neurons are interspersed among the epithelial cells of both tissue layers.

Ctenophores, the most primitive extant neural animals, possess a simple, radially symmetrical body plan organized along an oral-aboral axis. Their internal cavity has a single opening for feeding and excretion at one end, while a sensory organ is located at the opposite end. They also feature eight external rows of comb plates, composed of thousands of linked cilia that beat rhythmically to propel the ctenophore through the water.

Although cnidarians similarly lack distinct respiratory, circulatory, or excretory organs, their bodies are composed of two cell layers surrounding a gastrovascular cavity. Their tentacles, arranged around the mouth, capture and ingest food. Unlike ctenophores, some cnidarians have developed an elaborate nerve ring with a degree of regionalization. For instance, cubozoan medusae possess rhopalia—neural centers composed of thousands of neurons associated with their complex eyes. However, a single integrative center or "brain" is still absent.

The emergence of a bilaterian body plan introduced a developmental structure organized along a sagittal plane. Several key innovations define bilaterians: the mesoderm as a third germ layer, a complex bilateral nervous system, a Hox gene cluster with at least seven members, a through-gut, an excretory system, numerous specialized cell types, and bilateral symmetry. This body plan facilitated the dual use of the coelomic cavity for both circulation and digestion. Even after these functions became distinct in groups such as nematodes and mollusks, many organisms alternated between circulatory and digestive dominance. Most bilaterians have a continuous gut extending from mouth to anus.

Having a distinct head and anterior end meant that bilaterians encountered stimuli, such as food, at the front of the body, encouraging the development of sensory organs in this region. This anatomical shift promoted cephalization, the concentration of nervous tissue at the anterior end, forming a new type of neural organization, with an integrative center.

Consequently, the bilaterian nervous system developed a central neural structure—a *brain*. Defining the minimal requirements for a brain remains a challenge, but one common criterion is a distinct nervous structure composed of specialized neuronal clusters (nuclei) connected by axon tracts (neuropil). These clusters process sensory information and coordinate bodily responses via muscle activation and neurosecretion. Here, the term "brain" refers to a well-defined concentration of densely connected neural cells, typically located in the head near sensory organs, serving as an integration and control center.

The bilaterian body plan enabled greater diversification and complexity in both body structures and nervous systems, which allowed the appearance of the first land animals. As with many major evolutionary innovations, the realization of these possibilities depended on additional factors and adaptations. Tracing early bilaterian evolution is challenging due to the rapid succession of mass extinctions and subsequent radiations of new animal forms. Shortly after bilaterians emerged, the Cambrian explosion occurred, marking the sudden appearance of nearly all major extant body plans.

The reasons for this rapid diversification remain debated. Environmental factors such as increased atmospheric oxygen and rising oceanic calcium levels may have created favorable conditions for animal life. Some scientists propose that ecological shifts—such as changes in food webs, competition, and predator-prey dynamics—spurred massive coevolution. Others emphasize the role of genetic and developmental innovations.

In this context, the emergence of the bilaterian body plan may itself have contributed to the rise of new developmental designs, provided that environmental conditions were favorable. The specific body plans that emerged during the Cambrian explosion represent variations within the broader bilaterian framework. This idea aligns with evidence suggesting that bilaterians existed before the Cambrian and that further organizational innovations continued to drive diversification well into the Ordovician period and beyond.

Regardless of the specific causes, the emergence of new body plans within the bilaterian framework significantly influenced the evolutionary trajectories of Cambrian fauna. The potential for increased nervous system complexity—including the evolution of cephalization—was largely shaped by the constraints and opportunities imposed by these novel body architectures. As Ginsburg and Jablonka have suggested, it is likely that during the Cambrian explosion, newly evolved animals with more complex brains developed new forms of associative learning. This, in turn, would have enabled the expression of more sophisticated behaviors.

Bilaterians evolved along two major nervous system pathways: protostomes, where neural cells form a ventral nerve cord, and deuterostomes, where neurons initially organize into a dorsal tube before centralizing near sensory organs in a process called encephalization. Among protostomes, nervous system evolution generally led to clusters of neural cells (ganglia) distributed along the body, while in deuterostomes—particularly vertebrates—neural organization became increasingly centralized. Nonetheless, both groups include cases of significant cephalization.

Despite these evolutionary trends, the majority of bilaterian lineages did not develop highly complex nervous systems. Only a tiny fraction of animal species evolved large, sophisticated brains. The question of why this occurred will be explored next.

6.3 The Evolutionary Path to Complex Brains

How did cephalization lead to the development of complex brains in different evolutionary lineages? While there is no universal agreement on what constitutes a "complex" brain, it is widely accepted that such structures have emerged in only a few bilaterian phyla—arthropods, mollusks, and chordates. The extensive research devoted to this question focuses primarily on vertebrates, particularly land vertebrates, as well as cephalopods (mollusks) and certain social insects (arthropods).

Philosopher Michael Trestman suggests that brain and cognitive complexity arose only in lineages that evolved "complex active bodies". More recently in collaboration with Sheryl Coombs, he has argued that this complexity in vertebrates, cephalopod mollusks, and arthropods is closely linked to the development of body, sensory, brain, and motor traits essential for advanced visual processing and visuomotor skills. These traits enabled a shift from reactive to proactive behavior and from slow, two-dimensional movement to rapid, three-dimensional motion. High-resolution eyes and laminated visual brain regions significantly increased both information power and processing demands. The emergence of these traits resulted from evolutionary transitions that combined various features in different ways, leading to distinct levels of complexity.

However, achieving truly complex behavior requires not only centralized brains but also significantly larger ones, composed of millions or even billions of neurons. The definition of a "big brain" remains debated, but comparative studies often assess cephalization by measuring brain volume relative to body size. Among animals, only cephalopods and most vertebrates possess brains containing dozens or even hundreds of millions of neurons. This pattern suggests a clear correlation between body size and brain size in these groups.

Michael Tomasello has characterized "higher degrees of complex behavior" as a form of "psychological agency," defined as the capacity for planning, decision-making, and executive control in response to changing circumstances. While the classification of certain social insects with relatively small brains (around 1 million neurons) as psychological agents remains debatable, cephalopods and vertebrates overwhelmingly dominate among animals exhibiting big brains along with these advanced behaviors.

6.3.1 The Emergence of Large Brains

Why have large brains evolved only in certain groups? Empirical evidence suggests that big brains capable of supporting advanced behavior exist only in relatively large bodies. While some arthropods, such as the Japanese spider crab (*Macrocheira kaempferi*) and the giant coconut crab (*Birgus latro*), reach considerable sizes, their brains remain small. *B. latro* has a slightly larger brain than the eusocial honeybee (*Apis mellifera*), whose brain contains only about 1 million neurons—far fewer than those found in cephalopods and vertebrates. This raises the question of whether the arthropod body plan imposes constraints on brain size, limiting its further complexity. Conversely, the original body plans of mollusks and chordates may have possessed features that facilitated the evolution of larger brains under specific conditions.

The emergence of large brains likely depends on multiple factors. A crucial aspect is to be embodied in animals capable to grow in size while maintaining flexibility, strength, and speed—an adaptation seen only in vertebrates and cephalopods. As previously mentioned, rapid and highly flexible movement is closely linked to high-resolution sensory processing. Active sensing requires fast and precise motor outputs, which in turn demand advanced neural processing and learning capabilities. This feedback loop creates selective pressures for brain reorganization and expansion.

6.3.2 Cephalopod Evolution: The Role of Motility

Fast movement has played a key role in cephalopod evolution. The internalization and reduction of the ancestral mollusk shell allowed for an expansion of the mantle cavity, which in turn enhanced jet propulsion. By drawing water into the mantle cavity and expelling it through the siphon, cephalopods generate thrust, enabling rapid and flexible movement. This structural adaptation also led to the development of an advanced circulatory system, with a closed vascular network and chambered hearts that efficiently distribute oxygen and nutrients.

A closed circulatory system provides multiple advantages. It allows for a more efficient distribution of resources tailored to different organs, enables selective regulation of blood flow, and ensures that transport molecules remain within vessels until needed. These features support homeostatic integration and enable sustained high levels of activity. As a result, cephalopods can execute rapid directional changes, escape predators by jetting into the air, and even glide over significant distances—such as the common hook squid (*Onychoteuthis banksii*), which has been observed gliding more than 40 m.

Interestingly, cephalopods and vertebrates share similar neurogenesis mechanisms, suggesting that the evolutionary transition to large brains in cephalopods may have been driven by selection for greater body flexibility, rapid motion, and high-resolution sensory perception. These adaptations may have been advantageous in competing with vertebrate fishes.

6.3.3 Vertebrate Evolution: The Endoskeleton Advantage

Vertebrates also developed an efficient infrastructure for rapid movement at larger sizes, but through entirely different mechanisms. Their closed circulatory system, coupled with an internal endoskeleton, provides a strong yet flexible framework for muscle attachment. Additionally, vertebrates possess a more centralized nervous system, which enhances coordination and control.

A key factor in vertebrate evolution was the emergence of multipotent neural crest cells. These migratory cells, unique to vertebrates, give rise to various structures, including most of the cranial peripheral nervous system and sensory organs. Innovations such as jaws, teeth, and vertebrae played a crucial role in the evolutionary and ecological success of jawed vertebrates, which now comprise 99.8% of all extant vertebrate species.

Thus, both vertebrates and cephalopods evolved large, complex brains and highly mobile bodies, which likely provided selective advantages that outweighed the energetic and organizational costs of maintaining such brains. However, the similarities end there. While cephalopods developed both complex brains and body structures in a mutually functional yet largely independent manner, vertebrates evolved a more integrated system in which brain and body co-evolved in a tightly interconnected feedback loop.

6.3.4 Divergent Paths in Cephalization

One key distinction between cephalopod and vertebrate cephalization lies in how their brains regulate physiological processes. While the cephalopod neuroendocrine system controls many essential functions, only vertebrates evolved a finely tuned brain-driven autonomic nervous system (ANS), which directly regulates critical homeostatic functions such as heart rate, blood pressure, respiration, and digestion. This control system, particularly advanced in mammals, ensures precise physiological regulation.

For instance, the ANS governs vascular contractility and wall tension. Sympathetic and parasympathetic nerve fibers innervate blood vessels, while endothelial factors modulate vascular responses. This "cross-talk" between the ANS and the circulatory system is a hallmark of vertebrate physiology and represents a fundamental difference from cephalopods.

The distinction between vertebrate and cephalopod brain organization ultimately stems from their different evolutionary histories. Cephalopods expanded their brains primarily to support high-resolution sensory organs, such as octopus' eyes, along with extremely plastic bodies capable of shape and color changes, and also flexible and fast-moving limbs. This expansion also enhanced learning and information processing. In contrast, vertebrates developed a brain-body relationship that enabled complex physiological regulation in large-bodied organisms, ensuring both brain functionality and a strong and plastic body motility, even greater than that of cephalopods.

Vertebrates evolved a feedback mechanism in which brain development shapes the body, and vice versa. This continuous interaction fosters the emergence of increasingly centralized and integrative neural structures. In contrast, cephalopod neural integration appears to occur at a much lower level. For example, despite having 500 million neurons, the octopus brain contains only about 40 million neurons in its central structure. The optic lobes contain another 120–180 million, but the majority—about 300 million—are distributed across the arms for autonomous limb control. In contrast, birds and mammals of equivalent brain size allocate most of their neural resources to centralized control.

These divergent evolutionary paths likely reflect the constraints and possibilities inherent in each body plan. Vertebrates, from their early origins, possessed mechanisms for fine-tuned control of internal organs, enabling the development of highly efficient, strong, and fast movement across aquatic, terrestrial, and aerial environments. This adaptability facilitated their evolutionary radiation, resulting in around 65,000 extant species. In contrast, cephalopods, despite their intelligence, remain confined to marine habitats, with only about 800 known species.

Ultimately, these differences highlight the role of body plan constraints in shaping the evolutionary trajectories of complex brains. While both cephalopods and vertebrates achieved advanced cognitive abilities, their paths to cephalization reflect the unique challenges and opportunities presented by their respective morphologies.

6.4 Conclusion: The Emergence of Mind in the Evolution of Animal Agency

It is reasonable to defend that some kind of "mind" gradually emerged and evolved independently in both cephalopods and vertebrates. In the animal world, the term "mind" generally refers to the source of phenomena such as representational perception, emotions, feelings, conscious experiences, and action planning. These capacities likely developed over time in species with increasingly large and complex brains.

The emergence of these cognitive abilities is widely regarded as a qualitative shift in brain function, giving rise to a higher level of behavior often described as "mental agency." This transition may correspond to the evolution of animals with large brains capable of exhibiting what Tomasello terms "psychological agency". In these species, mental agency would have enabled advanced behaviors such as forming predictive associations between cues and controllable actions, as well as the ability to assess and re-evaluate action outcomes based on experience.

However, to consider the appearance of mind as a qualitative evolutionary change, we must also examine the distinct and far-reaching consequences it has triggered. If the emergence of mind is an emergent and qualitatively significant novelty, a naturalistic explanation must demonstrate that it represents a fundamental shift within the framework of evolutionary biology.

In the final part of this chapter, we have attempted to provide such an explanation. Without directly linking any specific neurophysiological structure to the emergence of mind, we have sought to clarify how and why larger brains evolved and why, in certain evolutionary lineages, these brains developed within organisms exhibiting complex forms of psychological agency. We have also explored the evolutionary impact of this development, showing that it initiated a new mode of evolution—one shaped by a distinctive form of brain embodiment. This process created a "virtuous causal cycle" in which brain, body, and behavior co-evolved in an accelerating feedback loop, driving ever-increasing complexity.

Bibliography

Arnellos A, Keijzer F (2019) Bodily complexity: integrated multicellular organizations for contraction-based motility. Front Physiol 10:1268. https://doi.org/10.3389/fphys.2019.01268

Arnellos A, Moreno A (2015) Multicellular agency: an organizational view. Biol Philos 30:333–357. https://doi.org/10.1007/s10539-015-9484-0

Arnellos A, Moreno A (2022) Cognitive functions are not reducible to biological ones: the case of minimal visual perception. Biol Philos 37:1–25. https://doi.org/10.1007/s10539-022-09860-3

Barandiaran X (2008) Mental life: a naturalized approach to the autonomy of cognitive agents. PhD Thesis, University of the Basque Country

Bechtel W, Bich L (2024) Eating and cognition in two animals without neurons: sponges and *Trichoplax*. Biol Theory. https://doi.org/10.1007/s13752-024-00464-6

Brunet T (2023) Cell contractility in early animal evolution. Curr Biol 33:R966–R985. https://doi.org/10.1016/j.cub.2023.07.054

Burkhardt P, Jékely G (2021) Evolution of synapses and neurotransmitter systems: the divide-and-conquer model for early neural cell-type evolution. Curr Op Neurobiol 71:127–138. https://doi.org/10.1016/j.conb.2021.11.002

Coombs S, Trestman M (2025) A multi-trait embodied framework for the evolution of brains and cognition across animal phyla. Behav Brain Sci Published online 2024:1–52. https://doi.org/10.1017/S0140525X24000335

Christensen W (2007) The evolutionary origins of volition. In: Ross D, Spurrett D, Kincaid H, Stephens L (eds) Distributed cognition and the will: individual volition and social context. MIT Press, Cambridge, pp 255–287

DiPaolo E, Bhurman T, Barandiaran X (2017) Sensorimotor life: an enactive proposal. Oxford University Press, Oxford

Ginsburg S, Jablonka E (2019) The evolution of the sensitive soul: learning and the origins of consciousness. MIT Press, Cambridge, MA. https://doi.org/10.7551/mitpress/11006.001.0001

Godfrey-Smith P (2020a) Metazoa: animal life and the birth of the mind. Collins, London

Godfrey-Smith P (2020b) Gradualism and the evolution of experience. Philos Top 48:201–220. https://www.jstor.org/stable/48628592

Jékely G (2021) The chemical brain hypothesis for the origin of nervous systems. Philos Trans R Soc Lond B Biol Sci 376:20190761. https://doi.org/10.1098/rstb.2019.0761

Keijzer F, van Duijn M, Lyon P (2013) What nervous systems do: early evolution, input-output, and the skin brain thesis. Adapt Behav 21:67–85. https://doi.org/10.1177/1059712312465330

Moreno A (2023) Some reflections on the evolution of conscious agents: the relevance of body plans. Biosemiotics 16:35–43. https://doi.org/10.1007/s12304-023-09525-y

Najle SR, Grau-Bové X, Elek A et al (2023) Stepwise emergence of the neuronal gene expression program in early animal evolution. Cell 186:4676–4693.e29. https://doi.org/10.1016/j.cell.2023.08.027

Newman SA, Benítez M, Bhat R et al (2025) Agency in the evolutionary transition to multicellularity. Quart Rev Biol 100:83–118. https://doi.org/10.1086/735964

Nguyen NM, Farge E (2024) Mechanical induction in metazoan development and evolution: from earliest multi-cellular organisms to modern animal embryos. Nat Commun 15:10695. https://doi.org/10.1038/s41467-024-55100-5

Niklas KJ, Newman SA (2020) The many roads to and from multicellularity. J Exp Bot 71:3247–3253. https://doi.org/10.1093/jxb/erz547

Rosslenbroich B (2009) The theory of increasing autonomy in evolution: a proposal for understanding macroevolutionary innovations. Biol Philos 24:623–644. https://doi.org/10.1007/s10539-009-9167-9

Ruiz-Mirazo K, Moreno A (2012) Autonomy in evolution: from minimal to complex life. Synthese 185:21–52. https://doi.org/10.1007/s11229-011-9874-z

Ruiz-Trillo I, Kin K, Casacuberta E (2023) The origin of metazoan multicellularity: a potential microbial black swan event. Annu Rev Microbiol 77:499–516. https://doi.org/10.1146/annurev-micro-032421-120023

Tomasello M (2022) The evolution of agency: behavioral organization from lizards to humans. MIT Press, Cambridge MA

Thompson E (2007) Mind in life: biology, phenomenology, and the sciences of mind. Harvard University Press, Cambridge, MA

Chapter 7
Epilogue

Abstract We examine the biological nature, evolution, and increasing complexity of agency, tracing its historical development and theoretical implications. The chapter explores how agency emerged as both a product of an organism's constitutive organization and a force driving its evolutionary continuity. The discussion addresses the defining characteristics of agency, highlighting its role in biological interactions, ecological networks, and the transition from simple to complex forms of life. Special attention is given to the relationship between motility, sensory systems, and constitutive development, emphasizing the evolutionary pressures that shaped complexification of agency in animals. Finally, we outline future research directions, particularly in synthetic biology and interdisciplinary collaborations, to further investigate agency's origins and transformations. By integrating insights from biology, philosophy, and artificial intelligence, this work aims to refine the conceptual framework of agency and inspire new empirical studies on its role in evolution.

Keywords Agency definition · Agency evolution · Sensorimotor integration · Synthetic biology

The epilogue of a book serves not only to present conclusions but also to assess its objectives and propose directions for future research. Our initial goal was to develop a new synthesis that explains the significance of *agency in biology*—how it originated, its function, its evolutionary trajectory across different branches of the tree of life, and why it has become more complex in certain lineages, and its consequences. Beyond offering a selective and synthetic account of agency's history, we aimed to generate new ideas that could inspire future scientific inquiry into this fundamental phenomenon and its role in life's evolution.

Some readers may have expected a greater focus on the evolution of highly complex forms of agency, particularly those associated with consciousness and language. As evolution has produced increasingly sophisticated modes of

Á. Moreno, J. Peretó, *An Evolutionary Story of Agency*, Foundations for Interdisciplinarity in the Life Sciences: Concise Monographs,
https://doi.org/10.1007/978-3-032-05644-3_7

agency—especially in lineages where complex brains have developed—agency itself has become the subject of new scientific disciplines, such as psychology and neuroscience. These fields introduce concepts like representation, imagination, emotion, feelings, and desires, often grouped under terms such as "mind" and "mental agency".

While numerous essays explore these advanced forms of agency, the origins and evolutionary development of agency in its various manifestations have received less attention. Yet, understanding the foundations of complex agency, including its cognitive aspects, requires an appreciation of its earlier evolutionary stages. This is why our focus has been primarily on the origins and initial evolutionary steps of agency.

With this in mind, it is now appropriate to revisit and refine our understanding of agency as a biological phenomenon.

7.1 What This Historical Unfolding of Agency Shows About Its Biological Nature

In this work, we have explored the phenomenon of agency through its historical expressions, examining how and why it emerged and what selective advantages it conferred, while assuming continuity and diversification in its evolution across different life forms. Our goal has been to contribute to a coherent conceptual framework for a comprehensive understanding of biological agency.

Agency in biology cannot be understood in isolation or without considering its historical context. Even in its simplest form, a natural agent presents numerous puzzles and paradoxical complexities. This makes the study of biological agency far more intricate than in disciplines like robotics or computer science. Unlike artificial forms of agency, biological agency must be explained without invoking external design or purpose, while also accounting for the extraordinary complexity of the living world, which introduces additional conceptual challenges in defining what constitutes a biological agent.

Biological agents arise from two fundamental conditions: a long and cumulative process of evolutionary development across countless individual lifespans and an embeddedness within broader ecological networks that sustain metabolic complementarities. They exist within an intricate multidimensional organization, shaped by interdependent systems operating at different spatial and temporal scales. Each component is present not only due to prior evolutionary steps and the influence of coexisting elements but also because of its own physiological, ecological, and evolutionary functions.

Moreover, biological agents frequently engage in intimate associations, where functions of individual entities are transferred to higher collective levels. Some extremely simple systems, such as viruses, act on their own behalf yet paradoxically rely on more complex organisms to do so. In other cases, collective forms of agency emerge from coordinated interactions, raising the question of whether such systems represent new integrated agents or merely the sum of individual actions. The biological world is thus filled with counterintuitive cases that challenge any straightforward definition of agency.

Ultimately, these complexities stem from the fundamental requirement that biological agents must autonomously construct themselves, reproduce and sustain ecological networks in which their functional interactions unfold across different temporal and spatial scales. Agents operate at microscopic timescales, while large ecological groups composed of similar agents function over macroscopic timescales, creating an intricate interplay of interdependent processes.

Rather than being a limitation, these challenges make the study of biological agency an invaluable opportunity to grasp the phenomenon in its full depth and complexity.

7.2 How to Define What Is Agency in Biology

Some readers may question how we can adopt a concept—agency—that lacks a widely accepted definition in the life sciences. This book is, in part, an attempt to address that concern. First, we acknowledge this issue and begin with a broad, commonly accepted notion of agency. Second, our goal has been to refine and elaborate this concept by tracing the evolutionary development of interactive behaviors in organisms.

The term "agency" in biology carries different meanings, and no single definition enjoys universal consensus. One might assume this issue could be resolved by establishing a strict definition that clearly delineates what qualifies as agency. However, defining it a priori does not provide a substantive argument for excluding alternative perspectives. As Alan Love aptly notes, "it might be tempting to think that the first step in the study of biological agency is to define clearly and unambiguously what it is. However, this is likely to immediately bring investigative efforts to a halt. There are many ways to construe agency that have been the subject of long discussions and no consensus definition currently exists." The idea is, therefore, to adopt a sufficiently generic starting concept to allow us to develop a piece of research whose results will help us to better understand the meaning of the agency phenomenon in biology.

There is no inherent reason to favor a broad or restrictive concept of agency. Some scholars consider long-term ecological impacts of populations, well-defined entities in ecosystems, or even viral pathogenic activity as forms of agency. Likewise, collective systems such as biofilms or eusocial insect societies are often classified as agents. To grasp the nature and significance of agency in biology, however, we must explore how and why this capacity has evolved differently across lineages and how certain animals developed highly complex forms of agency—psychological, conscious, or mental—including human agency.

Not all perspectives on agency are equally useful for understanding its evolutionary diversity. Our approach has been to examine the mechanisms that enable or constrain the diversification and increasing complexity of goal-directed interactions throughout evolution. Ultimately, our purpose has been to start from a widely accepted idea of agency and, through this research, attempt to develop a more elaborate concept of agency that provides explanatory power, helping to clarify the evolutionary processes that gave rise to its most advanced forms.

This approach may seem restrictive or overly demanding. However, we adopt it precisely to understand agency as an evolutionary continuum, tracing its development from its most rudimentary manifestations. This does not preclude recognizing forms of agency in non-organismic systems whose actions contribute to their collective self-maintenance. The challenge lies in avoiding a definition so broad that it obscures the complexity of agency's evolutionary history.

We did not begin with a rigid definition. In Chap. 2, we used "agency" synonymously with "functional interactive actions"—actions that affect the environment in ways beneficial to the system that initiates them. Similarly, we defined "agent" as any biological (or infra-biological) system capable of generating actions that are, directly or indirectly, functional for itself.

More specifically, we have begun by broadly defining agency as the adaptive interactions exerted by individuated biological systems on their environment, eventually accepting as candidates of agents any proto-organismic, or supra-organismic entity capable of achieving this type of flexible interactive loops. But, once again, we insist that this is only an initial statement, which must be reviewed throughout its evolutionary history. This framing situates agency within an evolutionary trajectory, where outward actions provide selective advantages that drive further evolution. Our choice of this working definition is pragmatic: it allows us to explore how agency evolved and displayed a diversity of complex forms.

From this perspective, non-organismic systems that contribute to collective self-maintenance can also exhibit agency, provided they are composed of individual systems that meet our criteria. The crucial point is that our definition must support an understanding of agency's evolution rather than hinder it.

Our strategy has been to trace the evolutionary processes that led to the clearest and most widely accepted forms of agency—those found in complex animals. We began by analyzing systems that developed and expanded their interactive capacities, focusing on those that initiated key evolutionary trajectories toward increasing complexity. This led us to examine the mechanisms underlying this specific evolutionary path. Based on this analysis, we now propose a refined understanding of agency and its role in life's history.

We define biological agency as *an adaptive interactive process performed by an individuated self-producing and reproducing system (an organism) that modifies its environment in ways that support its maintenance, either as an individual or as part of a specific lineage*. This definition establishes a clear organizational distinction and causal complementarity between the agent's constitutive organization, which generates its actions, and the interactive processes that, if successful, ensure the agent's continued existence.

This circular relationship necessitates identifying the agent as both the initiator and the beneficiary of its own actions. It also underscores the requirement of autonomy—where the agent sustains itself through its actions—as a naturalistic basis for understanding goal-directed behavior. In this sense, actions are "intended" to achieve a goal, meaning they contribute to maintaining the agent's identity; if they fail, the agent will cease to exist.

We adopt this definition not only because it captures the fundamental nature of biological agency but also because it provides a framework for tracing its evolutionary trajectory. It serves as a foundation for understanding the multiple evolutionary pathways that have led to the increasing diversification and complexity of agency over time.

Far from halting investigation, we think that this definition serves as a foundation for further exploration. As we have seen, the components of this definition have become increasingly relevant throughout evolution, shaping the distinct paths agency has taken. The following sections, particularly Sect. 7.4, will further clarify what the evolutionary history of agency reveals about its nature.

7.3 What Does the Evolution of Agency Show?

In every stage of evolution, agency has driven and shaped distinct evolutionary trajectories. Agency has been exploring different forms of organization that have opened up at each stage, and in each case the type of agency—determined by the ecological trophic niche of the organism: autotrophic primary producers, heterotrophic decomposers or consumers, etc.—has determined in turn a specific set of mechanisms that has been under selection during their evolutionary paths.

For example, the six forms of eukaryotic complex multicellularity have emerged from distinct organizational and ecological contexts. To explain these diverse evolutionary paths, we have considered the ecological circumstances within which each form of multicellularity evolved. Multicellular algae and plants evolved as autotrophic, sessile organisms, relying on photosynthesis and remaining fixed in one location. Fungi are heterotrophic organisms that have adapted to a sessile lifestyle, secreting enzymes and decomposing external organic matter before absorbing it into their cells. Metazoans, by contrast, evolved as heterotrophic, motile organisms that actively search for and internally digest organic material.

These fundamental ecological distinctions have significantly influenced how each lineage developed its multicellular organization leading to major variations in organismal complexity across evolutionary history. But the ecological circumstances in which each form of multicellularity has evolved have not determined their fate, because these multicellular organisms, with their actions, are themselves the creators of new ecological niches. Since the range of conditions necessary for the persistence of a species—where it lives, what it eats, how it avoids predators or on what other species it depends—is also to a large extent the result of its past evolutionary activity and of that of the other ones with which this species interacts. Often, this includes a huge number of species, which together constitute an ecosystem. The behaviors of each agent are fundamental not only in each niche construction, but also in the whole ecosystem construction. For example, dung beetles play a crucial role—both ecologically and economically—in the decomposition of large herbivore dung; many plants depend on insects for their reproduction, whereas others, rely on moving animals that eat their seeds for dissemination; and fungal-plant symbiosis have expanded their respective possibilities.

There are some forms of agency—in particular, most forms of agency linked to motility at multicellular size—which, once available, have favored a selection increasing body complexification. Throughout these pages we have tried to explain how. The answer is complex: multicellularity has enabled new forms of organization, and some of these have opened up new possibilities for agency, which, once they appeared, have in turn allowed selection to drive them in new evolutionary directions.

These directions have led to very specific new body plans, whose evolutionary success was linked to new forms of sensory-motor action, which in turn have opened new niches. Both factors have been generating a reciprocal causal action: conquer of new niches based on new forms of motile agency whose effective deployment is very demanding in terms of internal organization. To understand this process of reciprocal action, it is essential to realize the immense diversity and complexity of behaviors that sensory-motor motility opens up on a multicellular scale. The opportunistic exploitation of this enormous space of new functional possibilities is what marks the evolutionary bias of morphospace.

Yet, a full understanding of the evolution of agency implies also to address the consequences of this process of complexification on the global history of life. In other words, which have been the effects of the complexification of agency in the evolution of biosphere.

Since the beginning of life, all living systems have been forming a global system that actively participates in its own continuity, through the geochemical cycles. The byproducts of life give rise to powerful bio-geochemical recycling of essential nutrients such that the availability of essential bioelements vastly exceeds that expected solely from abiotic influxes. Thus, by modulating its environment for its own persistence and proliferation, the phenomenon of life has persisted during 4000 my.

Early life on Earth likely had a very small biomass composed of microorganisms. But during the history of life the biosphere has changed, sometimes due to abiotic causes, and other times due to evolutionary events. Thus, certain evolutionary novelties have had a deep impact on environmental conditions and biogeochemical cycles. For example, the oxygenation of the planet, produced by cyanobacteria capable of chlorophyl-based oxygenic photosynthesis, probably originated a mass extinction but, in turn, it opened the door of novel biological innovation and diversification.

The role of complex organisms and their forms of agency are fundamental for understanding the history of biosphere. The emergence and diversification of complex life has also altered the global biosphere, impacting on environmental conditions and biogeochemical cycles. Complex life has played a key role in shaping ecosystems and driving biodiversity by occupying diverse niches and influencing ecological processes, including increase in global biomass production. Complex life and biodiversity are intertwined, with complex life driving and being influenced by biodiversity. For example, the appearance of new forms of complexity during the history of life has regularly undermined the habitable conditions initially allowing it to flourish and that biology destabilizes its environment (massive extinctions) at least as frequently as it induces any form of homeostasis within it. The biosphere has evolved in an irreversible way, enacting new forms of global ecological maintenance built on the new forms of life that constitute them historically, because trophic chains involving complex behavior's organisms, and creating new niches are fundamental pieces for the maintenance of these new forms of biosphere. All this further highlights the need to understand why agency has played such a remarkable role in the history of the complexification of life.

7.4 What the Historical Unfolding of Agency Discloses

Life has developed a vast array of adaptive strategies to cope with changing environments. Beyond the long-term evolutionary changes that shape populations over generations, organisms possess diverse mechanisms for adapting within their own lifetimes. However, what makes the agency unique, and at the same time amenable to evolution, is that it operates through action-detection loops, where actions are reversible processes continuously modulated by sensory feedback. For agency to be effective, sensory information must be updated as frequently as necessary to guide actions toward their intended goal. In other words, successful agency relies on rapid adjustments, ensuring that actions are corrected swiftly enough to achieve their desired outcome. This makes agency a specialized, rapid-response adaptive mechanism, consisting of repeatable and reversible processes.

This distinction is crucial for understanding the different types of agency and their evolutionary consequences. In secretion-based agency, once a component is released, the agent has no strict control over its effects, even if the latter are eventually there, and detected by the system, closing the loop. By contrast, motile agency allows for continuous action-detection control, linking movement with the precision and speed of sensory feedback. This connection drives the refinement of sensory organs, the accuracy of motor responses, and the increased capacity for processing sensorimotor information. Only motility enables an agent to regulate the speed and effectiveness of its actions, opening the possibility for new behaviors—such as capturing a fast-moving prey—that require highly efficient action-detection loops.

Motile agency also has an essential relationship with the size of the agent. As organisms increase in size, they gain new interactive possibilities, but these require the ability to execute fast, coordinated movements on a larger scale. Unlike secretory agency, which remains relatively unaffected by size, motile agency demands a more complex support system as body size increases. This is particularly evident in multicellular metazoans, which interact with a macroscopic world through perceptual constructions, such as visual images. In contrast, sessile multicellular organisms like fungi and plants lack the capacity to perceive macroscopic objects in the same way. Even in cases like the Venus flytrap plant, which captures insects, the size of the prey is largely irrelevant to the plant's detector-response mechanism.

These distinctions explain why the increasing complexity of agency has primarily occurred through motility. The differences between motile and secretory agency account for the co-evolution of sensory organs, effectors, and the information-processing systems that connect them. This, in turn, reveals why the choice between a sessile or motile lifestyle has profoundly influenced evolutionary pathways. Among organisms that have increased in size, those that developed motility-based life forms have also undergone significant internal reorganization to support this mode of life, experiencing selective pressures distinct from those shaping sessile organisms.

By examining how agency has evolved and how its organizational demands change with increasing body size, we can better understand the connection between the minimal concept of agency—introduced in Chap. 2—and its more complex, evolved forms. This perspective allows us to trace the development of agency from its simplest manifestations to the conditions necessary for its evolution toward greater complexity—a path that inevitably involves motility. It also enables us to assess the concept of agency we have developed throughout this book.

Our concept of agency is not intended to be exclusive, nor does it dismiss broader definitions. However, many alternative perspectives treat biological agency as a universal feature of all organisms, influencing evolutionary trajectories in a uniform way, regardless of their complexity. DiFrisco and Gawne have strongly criticized such views, even questioning the overall usefulness of the agency framework in the study of life. In these pages, we have argued that this conclusion is misguided. Our work not only refines the theoretical foundations of the notion of agency but also supports empirical progress through experimental research within the life sciences. The approach developed in this book emphasizes that agency—a capacity of individuated forms of life, embodied in specific mechanisms—evolves, and that different forms of agency shape evolutionary lineages in distinct and transformative ways.

7.5 Concluding Remarks: Future Directions and Prospects

Before concluding, we would like to propose avenues for future research. What might be the utility of this book? In recent years, the question of agency has sparked increasing interest and debate in both biology and philosophy. This work was aimed not only to contribute to that discussion but also to help build a more unified theoretical framework in the life sciences.

The concept of agency has also entered the long-standing debate between a gene-centered biology and one focused on the organism. Several authors—most recently Alfonso Martínez Arias, Philip Ball, and Paul Nurse—have emphasized the need to "rediscover" the cell. Rather than adopting an anthropocentric notion of agency, evoked by intentional human actions, we have proposed investigating its origins in the adaptive protocells of the distant past and tracing the tree of life to identify the pathways through which agency has become more complex. There is no agency below the level of the cell, regardless of how intricate a chemical system may be. The cell is both the minimal unit of life and the minimal unit of agency. Our work may therefore contribute to the ongoing construction of a systems biology that places the organism, rather than its molecular components alone, at the center of explanation.

By offering new conceptual perspectives, we hope to inspire fresh lines of inquiry, particularly in evolutionary biology, where the search for empirical evidence is most productive when guided by novel theoretical insights. While theoretical reflection is essential, science cannot advance without empirical validation. Theories must be grounded in experiments and observations.

One of the most promising avenues for studying agency experimentally is synthetic biology. In the context of origin-of-life research, synthetic biology combines computational modeling with laboratory experimentation to create artificially designed protocells capable of autonomous behavior. Computational simulations further allow researchers to explore how populations of primitive protocells might have evolved into the earliest adaptive agents.

As Ricard Solé has suggested, every major evolutionary transition may have an artificial counterpart. Particularly relevant to our discussion is the development of synthetic multicellular systems that exhibit autonomous action. For instance, biohybrid robots or "bio-bots" are composed of embryonic cells that have been artificially evolved to form coherent multicellular entities capable of coordinated movement. Such research allows scientists to explore entirely new forms of multicellular agency—entities that have never existed in natural evolutionary history. By integrating novel inter- and intra-cellular constraints into their design, artificial multicellular agents may provide deeper insights into the role of cell motility in the origin of animals, shedding light on the mechanisms driving agency transformations and increasing complexity throughout evolution.

Ultimately, progress in understanding agency will require collaborative efforts across multiple disciplines. Synthetic biology, microbiology, evolutionary and developmental biology, ethology, robotics, and artificial intelligence must converge to advance future research programs in the life sciences. Achieving this goal will depend on fostering interdisciplinary collaboration, particularly by integrating scientific and philosophical approaches into a more comprehensive and cohesive framework.

Bibliography

Ball P (2024) How life works. A user's guide to the new biology. Picador, London

Ball P (2025) Nobel Laureate, Paul Nurse, on the future of biology. Meanings of life project. Marginalia Review of Books. https://www.marginaliareviewofbooks.com/post/paul-nurse-future-biology. Accessed 23 Apr 2025

Bich L (2024) Organisational teleology 2.0: grounding biological purposiveness in regulatory control. Ratio 37:327–340. https://doi.org/10.1111/rati.12405

Blackiston D, Lederer E, Kriegman S (2021) A cellular platform for the development of synthetic living machines. Sci Robot 6:eabf1571. https://doi.org/10.1126/scirobotics.abf1571

DiFrisco J, Gawne R (2025) Biological agency: a concept without a research program. J Evol Biol 38:143–156. https://doi.org/10.1093/jeb/voae153

Gumuskaya G et al (2024) Motile living biobots self-construct from adult human somatic progenitor seed cells. Adv Sci 11:e2303575. https://doi.org/10.1002/advs.202303575

Kauffman SA (2000) Investigations. Oxford University Press, UK

Love A (2023) Reflections on the study of biological agency and its evolution. Spon Gen 11:1–7. https://doi.org/10.4245/spongen.v11i1.13757

Martínez Arias A (2023) The master builder. How the new science of the cell is rewriting the story of life. Basic Books, New York

Mitchell KJ (2023) Free agents: how evolution gave us free will. Princeton University Press, Princeton

Moreno A, Etxeberria A (2005) Agency in natural and artificial systems. Artif Life 11:161–176. https://doi.org/10.1162/1064546053278919

Moreno A (2018) On minimal autonomous agency: natural and artificial. Compl Syst 27:289–313. https://doi.org/10.25088/ComplexSystems.27.3.289

Rosslenbroich B, Kümmell S, Bembé B (2024) Agency as an inherent property of living organisms. Biol Theor 19:224–236. https://doi.org/10.1007/s13752-024-00471-7

Solé R (2016) Synthetic transitions: towards a new synthesis. Philo Trans R Soc B Biol Sci 371:20150438. https://doi.org/10.1098/rstb.2015.0438

Solé R, Conde-Pueyo N, Pla-Mauri J et al (2024) Open problems in synthetic multicellularity. NPJ Syst Biol Appl 10:15. https://doi.org/10.1038/s41540-024-00477-8

Sultan SE, Moczek AP, Walsh D (2022) Bridging the explanatory gaps: what can we learn from a biological agency perspective? Bioessays 44:e2100185. https://doi.org/10.1002/bies.202100185

MIX
Papier aus verantwortungsvollen Quellen
Paper from responsible sources
FSC® C105338

If you have any concerns about our products,
you can contact us on
ProductSafety@springernature.com

In case Publisher is established outside the EU,
the EU authorized representative is:
Springer Nature Customer Service Center GmbH
Europaplatz 3, 69115 Heidelberg, Germany

Printed by Libri Plureos GmbH
in Hamburg, Germany